SILENT BUT DEADLY

More Homemade Silencers from Hayduke the Master

George Hayduke

Paladin Press
Boulder, Colorado

Also by George Hayduke:

Get Even: The Complete Books of Dirty Tricks
Get Even 2: More Dirty Tricks from the Master of Revenge
Hardcore Hayduke: More Down-and-Dirty Revenge Techniques
The Hayduke Silencer Book
Kickass! More Mayhem from the Master of Malice
Make 'em Pay! Ultimate Revenge Techniques from the Master Trickster
Make My Day! Hayduke's Best Revenge Techniques for the Punks in Your Life
Payback! Advanced Backstabbing and Mudslinging Techniques
Righteous Revenge: Getting Down to Getting Even
Screw Unto Others: Revenge Tactics for All Occasions
Sweet Revenge: A Serious Guide to Retribution
Up Yours! Guide to Advanced Revenge Techniques

Silent But Deadly: More Homemade Silencers from Hayduke the Master
by George Hayduke

ISBN 0-87364-811-0
Printed in the United States of America

Published by Paladin Press, a division of
Paladin Enterprises, Inc., P.O. Box 1307,
Boulder, Colorado 80306, USA.
(303) 443-7250

Direct inquiries and/or orders to the above address.

Contents

Introduction

A FEW TOO MANY WORDS FROM OLD HIMSELF

Tom Hritz is a professional curmudgeon and agitator whose cover includes writing a newspaper column in that provincial rust-belt antiquity known as Pittsburgh. Friend Tom's answer to gang violence and drive-by shooting was to issue the cracked-out kiddies silencers for their guns so they wouldn't disturb the good burghers guzzling their Iron City, a locally brewed holy water.

Tongue pushed firmly through his cheek, Tom had voiced an idea that was as realistic as most of the gun control ideas offered rabidly by his tooth-fairy-biased editors and their equally head-up-the-ass journalistic counterparts.

By contrast, one rare breath of journalistically honest fresh air came from the mind of social critic Rhyging Solipist, who wrote, "The major role of a firearm silencer is to greatly reduce the loud, abusive blast of a gunshot. I should think environmentalists, greenists, and other glass-ceiling liberals would be pleased about this cacophony-challenged technology."

Recall all the media fuss about "spud guns" during the strange summer of '93? Concerned law enforcement REMFs and their zealot-driven media partners were in anguish about a horrible new weapon revolutionizing gang warfare and allowing towel-headed international terrorists to seize lesser downtown Newark.

Actually, several Texas funsters were arrested, fined, and incarcerated for "being in the possession of a prohibited weapon" early in 1994. Their crime? They were silly-testing an inoffensive compressed air tube that propelled potatoes about 100 feet at a very low rate of slow.

The spud gun is the reinvention of the old Polish cannon, a fairly harmless toy that nondork guy kids "discover" about the age of 13. A Polish cannon, or spud gun (its name of choice among media harpies), is a pipe or other container 2 to 4 feet long "firing" a projectile that's propelled usually by ignited lighter fluid. In the old days, rich guys fired tennis balls while we poorer kids used hard green tomatoes in this silent blaster.

Media-hyped police REMFs in California said these guns were as dangerous as real firearms—totally silent and could fire a potato at 1,000 feet per second. Were there drive-by potatoings in Quayle, California?

Could we ban spud guns nationally? Possession of an unregistered spud gun is a felony fall in California and Texas as this is written. No shit, it is. Perhaps a McBrady Bill with a five-day waiting period? Why haven't our anti-spud media mavens editorially harrumphed about banning assault spuds?

Speaking of media, like being drug- and AIDS-free, I'd like to proudly proclaim here that I am not now, and have never been, a media maven. But, if I could offer some advice anyway (what the hell, they always give it to us), perhaps silencers could be applied to certain of the journalism ilk as readily as they are to firearms, all for the nobility of environmental greening, of course.

So, as a fellow social critic, I agree with Rhyging that silencers are useful in that they make great contributions to a saner, quieter, more pastoral way of life. And they don't cause scary diseases! They also interest a lot of folks, obviously, because my first book on the topic generated a lot of new friends with new ideas for simple silencer designs, i.e., the kind you can build at home.

Introduction

Despite the pucker-tightening rhetoric of the antigun assholes, there are very civilized uses for silencers. For example, during a recent Ladies' Golf Classic held in Birmingham, Alabama, the county sheriff's deputies provided course security using silencer-equipped rifles to dispatch snakes on the fairways, thus avoiding the noise that had disturbed play previously.

OK, there are times you want noise, though. For example, Ron Birdbuster was telling me how he deals with deer jacklighters and poachers out at his place.

"I have an old black powder cannon operated by remote control up on the hill about 200 yards from the road. When one of those assholes starts night hunting in my field I touch off the cannon with one switch . . . KAAABOOOOM!!! with all the noise and fire.

"I hit the second switch, which sets off an M80 wired into a half pint of gasoline down at the bottom of the hill, about 30 feet from the road. It's pretty impressive. But so are the rubber paths laid on the road by the messed-pants poachers."

Then there are the silent times when your goal would be compromised by noise. For example, you want to do some target shooting or can busting but your kid's asleep or the neighbors freak at the sound of gunfire. Solution? Fasten a silencer on your firearm.

My friend Fred Sykes was a suburban dweller and loved to garden. Grackles used to hold seed buffets in his garden. Fred had mouth-foaming antigun neighbors who called the police if he even talked about firing blanks to scare the birds. Solution? Fred bought a very efficient silencer for his .22 rifle.

Laughing through that evil slit in his face that he called a smile, Fred asked, "But why the hell did I buy a silencer when I could have built one?"

Indeed, why buy when you can build?

You *can* build a silencer for your own gun. It's all very legal if you do it properly. And that's the next section of this book—how to do the legal stuff easily and properly. Please pay attention to the rules. Since the '90s, when the former Soviet Union lost its paranoid dictatorship mentality, U.S. and Canadian government officials seem to have found it.

People who should have gone to prison for life over what they made happen in Waco, Texas, now own the socialist dictatorship in the United States. People should have gone to prison for life for what they did to Randy Weaver and his family in Idaho, too. If all the government assholes who have trampled our constitutional rights in the past three decades were put into prison, the District of Columbia would be inhabited only by petty thieves and drug dealers.

But, as Bob, that great talk show host from WIBC radio in Indianapolis, told me, I should change my literary credo from "DON'T GET MAD" to "GET ANGRY." I like the sound and the flow of that modified motto, but, there are advantages to the bad guys thinking you are a tad mad, as well as angry.

As my friend Thomas says quietly, "May Thomas Jefferson rest in peace, or at least walk side by side with Jesus if the Second Coming is allowed to happen as part of the New World Order."

Meanwhile, back at the gunshop, Anton Chekov was not necessarily correct, or even literal, when he wrote his famous dictum that if you show the audience a gun in the first act, it has to go off in the last.

Which means that you're wondering why this book's title, SBD, which is the universal acronym for a fart, is just that, our title. If you think about that, it will make less than perfect sense, which is why Karen, my personal editor, did it. And, as a personal friend of the Queen of Vulgarity, I feel she has the perfect right to do so.

In any case, play by the rules or you'll discover that old Berlin joke that Bernard Sampson shared with me many years ago: Glasnost is trying to escape over the Berlin Wall and getting shot with a silenced machine gun.

Introduction

As you enjoy this new home-built silencer book, please recall what George Hayduke told P.J. O'Rourke the last time they shared some hot chocolate in Miami: "I do not exist, yet I appear!"

—George Hayduke
The Research Institute of Precise Machines
Klimovsk, Russia
4 July 1994

1

Play Legal or Pay . . . Big Time!

His Lordship and Lady Sheila sailed on the QE2 to London, heading for their estate in the Cotswolds. Two days out of New York they were sitting on the top deck enjoying a clear, romantic night. The Atlantic was glassy and calm, like a gorgeous moonlit lake, a 360-degree uninterrupted view on all horizons.

Turning to his Lordship, Lady Sheila said with romantic wonder, "Oh, darling, look at all that water out there."

Broadening his Cheshire smile, his Lordship replied, "Yeah, and that's just the top part."

Which, in a far less innocent way, is the way our rulers view the rest of us here in Dictatorship Amerika.

The power elite in their corporate towers, seat of government (SOG) cartels, and law firms look over feudal America as their national plantation controlled from The Big White House and its attached Halls of Congress in the District of Columbia. As the workers and the slaves, we little folks are losing more and more freedom each day.

Using the narcotic smoke and mirrors of mass media, the power elite scare our freedoms out of us with their choreographed propaganda tactics about drugs, street gangs, gun control = violence/crime control, international terrorists, revolution, insurrection, and on and on into George Orwell's worst waking nightmare now come alive and true.

And, being pliantly hypnotized by the televised narcotic of power-elite propaganda, you slaves and workers out there in Plantation Amerika huddle closer to your electronic father and eagerly toss away your freedoms in return for lies and promises.

The Second Amendment is one of those freedoms that is about gone. It's too late for compromise and polemic. When the power elite finally trade enough lies to complete their capture of the Second Amendment, it will all be over. So, if you want to build a legal silencer for your firearms, do it now, while you still are able to get old Massa's approval from SOG.

Here's how it works.

First, you cannot acquire, assemble, manufacture, possess, purchase, or receive any silencer that is not legally registered with the federal government. To register this silencer, you must survive tedious bureaucracy and patronization. This involves getting a mug shot and fingerprints and filling out some federal cards and forms that require a local law enforcement signature.

Before you even get around to the federal forms, you will need to learn if silencers are legal in your local area, state, region, etc. A quick call to your nearest BATF office will settle that matter quickly. Or, you can call a nearby Class III dealer . . . more on that later.

To help you accomplish this task, I will explain these forms to you, tell you how to complete them, then walk you through the process. I have also reproduced copies of these forms in this book on pages 9 through 17.

Play Legal or Pay . . . Big Time!

DEPARTMENT OF THE TREASURY – BUREAU OF ALCOHOL, TOBACCO AND FIREARMS

APPLICATION TO MAKE AND REGISTER A FIREARM

DETACH THIS SHEET BEFORE COMPLETING FORM

INSTRUCTIONS

1. DEFINITIONS.

a. FIREARM. The term "firearm" means: (1) a shotgun having a barrel or barrels of less than 18 inches in length; (2) a weapon made from a shotgun if such weapon as modified has an overall length of less than 26 inches or a barrel or barrels of less than 18 inches in length; (3) a rifle having a barrel or barrels of less than 16 inches in length; (4) a weapon made from a rifle if such weapon as modified has an overall length of less than 26 inches or a barrel or barrels of less than 16 inches in length; (5) any other weapon, as defined in b. below; (6) a machine gun; (7) a muffler or a silencer for any firearm whether or not such firearm is included within this definition; and (8) a destructive device. The term "firearm" shall not include an antique firearm or any device (other than a machine gun or destructive device) which, although designed as a weapon, the Director, Bureau of Alcohol, Tobacco and Firearms, or authorized delegate finds by reason of the date of its manufacture, value, design, and other characteristics is primarily a collector's item and is not likely to be used as a weapon.

b. ANY OTHER WEAPON. The term "any other weapon" means any weapon or device capable of being concealed on the person from which a shot can be discharged through the energy of an explosive, a pistol or revolver having a barrel with a smooth bore designed or redesigned to fire a fixed shotgun shell, weapons with combination shotgun and rifle barrels 12 inches or more, less than 18 inches in length, from which only a single discharge can be made from either barrel without manual reloading, and shall include any such weapon which may be readily restored to fire. Such term shall not include a pistol or a revolver having a rifled bore, or rifled bores, or weapons designed, made, or intended to be fired from the shoulder and not capable of firing fixed ammunition.

c. PERSON. The term "person" means any individual, company, corporation, association, firm, partnership, joint stock company, trust or society.

2. PREPARATION OF APPLICATION TO MAKE AND REGISTER A FIREARM, AND PAYMENT OF TAX, WHERE REQUIRED.

a. As provided in §5822 of the National Firearms Act, every person [other than a licensed manufacturer who has also paid the required special (occupational) tax to manufacture NFA weapons] seeking to make a firearm must complete, in duplicate, a separate application on this form for each firearm. The applicant maker must furnish all the information called for on this application form.

b. The applicant must present this form to the law enforcement agency having jurisdiction in his area of residence (Chief of Police, Sheriff, etc.) for completion of the Law Enforcement Certification on the back of the form. If the applicant is other than an individual, the trade name should be entered as "name of maker."

c. If the applicant is an individual (including a licensed collector), an unmounted photograph (2" x 2" taken within the past year) must be affixed in the indicated space on both copies of the form, and completed FBI Form FD-258, Fingerprint Card, must be submitted in duplicate. The prints should be taken by someone qualified to do so and must be clear, unsmudged and classifiable. In addition, the person taking the fingerprints must enter the identification data regarding the individual maker and must complete the Fingerprint Cards by signing as the person taking the fingerprints.

d. All required signatures must be entered in ink. It is preferred that the form be prepared by use of a typewriter, using carbon paper to make an exact duplicate. Pen and ink may be used, but under no circumstances will a form filled in by use of a lead pencil be accepted. The signature on both copies *must* be an original. Photocopies, other facsimiles, or carbon copy signatures are not acceptable. All changes made on the form must be initialed and dated by the applicant.

e. Unless the making of the firearm is tax exempt (see instruction 4 below), a $200 making tax must be paid (Title 26 U.S.C., Chapter 53, §5821(a)).

f. Item 4a. of the form should clearly indicate if the parts of a receiver which had been destroyed in accordance with Department of Defense (DOD) demilitarization standards will be used to "make" the receiver of the firearm. The name of the manufacturer of the original receiver should be shown, as well as any serial number appearing on such receiver. Please note that a machinegun receiver which has not been destroyed according to current DOD demilitarization standards may be classified as a National Firearms Act (NFA) weapon in and of itself. It is unlawful to possess an NFA firearm which is not registered to the possessor and the weapon is subject to the seizure and forfeiture provisions of the law.

g. It is suggested that the Serial Number (item 4g on the face of the form) contain at least four digits, preceded by the maker's initials.

h. If any questions arise concerning the preparation of the form, contact the NFA Branch, Bureau of Alcohol, Tobacco and Firearms, Washington, DC 20226 or the nearest ATF office.

3. DISPOSITION OF APPLICATION TO MAKE AND REGISTER A FIREARM.

The applicant will forward both copies of the form to the NFA Branch, Bureau of Alcohol, Tobacco and Firearms, Washington, DC 20226, with a $200 check or money order (where required) made payable to the Department of the Treasury. If approved, the original of the approved form will be returned to the applicant and ATF will retain the duplicate. Approval by ATF will effect registration of the firearm to the applicant. The applicant shall not, under any circumstances, make the firearm until the approved form, with the "National Firearms Act" stamp attached, is received. The form must be retained by the applicant and be available at all times for inspection by Government officers until such time as the firearm may later be transferred after approval by this office. If the application is disapproved, the original and any accompanying check or money order will be returned to the applicant with the reason for disapproval.

4. EXEMPTIONS FROM PAYMENT OF TAX.

a. Under the provisions of §§5852 and 5853, National Firearms Act, firearms, as defined in 1a. and 1b., may be made by any person without payment of the making tax when made by, or on behalf of: (1) the United States, or any department, independent establishment, or agency thereof; or (2) any State or possession of the United States, or any political subdivision thereof, or any official police organization of such a government entity engaged in criminal investigations. The maker must apply on this form and obtain the approval of the Director before making the firearm. Documentation that the firearm is being made for a government entity must accompany the application. A U.S. Government Contract number or a State or local government purchase order would be acceptable documentation. Upon receipt of the approved ATF Form 1, and after the firearm has been made, the maker must apply on ATF Form 5 for the tax exempt transfer of the firearm to the government entity for whom it was made.

b. A manufacturer who has paid special (occupational) tax to manufacture firearms may make the kind of firearm that he is qualified to manufacture without payment of the making tax and he is not required to file this application form. However, the qualified manufacturer must report and register each firearm made by filing ATF Form 2, Notice of Firearms Manufactured or Imported, with the Director, immediately after manufacturing the firearm.

(Continued on reverse)

ATF FORM 1 (7560.1) (4-82)

This is the BATF Form 1, which is used to apply for permission to build and register a silencer. Filed in duplicate, it must be accompanied by two fingerprint cards and a $200 check. You must get a Form 1 approved, signed, and returned *before* you begin work on your silencer. (Courtesy of BATF)

INSTRUCTIONS (Continued)

5. PERSONS PROHIBITED FROM MAKING A FIREARM.

Section 5822 of the National Firearms Act requires that the application to make a firearm be denied if the making or possession of the firearm would place the person making the firearm in violation of law. The term "law" in this statute includes Federal laws as well as State statutes and local ordinances applicable to the locality where the transferee resides. Under Title VII of Public Law 90-351, as amended, (18 U.S.C., Appendix, 1201-1203), the possession "in commerce or affecting commerce" of a firearm, including an unserviceable firearm which has a frame or receiver is unlawful when possessed by any person who:

(1) has been convicted by a court of the United States or of a State or any political subdivision thereof of a felony, or

(2) has been discharged from the Armed Forces under dishonorable conditions, or,

(3) has been adjudged by a court of the United States or of a State or any political subdivision thereof of being mentally incompetent, or

(4) having been a citizen of the United States, has renounced his citizenship, or

(5) being an alien is illegally or unlawfully in the United States.

6. PENALTIES.

Any person who violates or fails to comply with any of the requirements of the National Firearms Act shall, upon conviction, be fined not more than $10,000 or be imprisoned for not more than 10 years, or both, in the discretion of the court. Moreover, any firearm involved in any violation of the provisions of the National Firearms Act or any regulations issued thereunder shall be subject to seizure and forfeiture. It is unlawful for any person to make or cause the making of a false entry on any application or record required by the National Firearms Act, knowing such entry to be false.

7. LATER TRANSFER OF THE FIREARM.

If the firearm is to be transferred later by the applicant, an application form covering the proposed transfer must be filed with the Director, Bureau of Alcohol, Tobacco and Firearms.

PRIVACY ACT INFORMATION

1. AUTHORITY. Solicitation of this information is made pursuant to the National Firearms Act (26 U.S.C. §5821 and 5822). Disclosure of this information by the applicant is mandatory for any person (other than a manufacturer qualified under the National Firearms Act) making a firearm as defined in the National Firearms Act.

2. PURPOSE. To verify payment of the tax imposed by 26 U.S.C. §5821; to determine that the making would not be in violation of law; and to effect registration of the firearm.

3. ROUTINE USES. The information will be used by ATF to make the determinations set forth in paragraph 2. In addition, to effect registration of the firearm, information as to the identification of the firearm, date of registration, and the identification and address of person entitled to possess the firearm will be entered into the National Firearms Registration and Transfer Record. No information obtained from an application, registration, or records required to be submitted by a natural person in order to comply with any provision of the National Firearms Act or regulations issued thereunder, shall, except in connection with prosecution or other action for furnishing false information, be used, directly or indirectly, as evidence against that person in any criminal proceeding with respect to a violation of law occurring prior to or concurrently with the filing of the application. The information from this application may only be disclosed to Federal authorities for purpose of prosecution for violation of the National Firearms Act.

4. EFFECTS OF NOT SUPPLYING INFORMATION REQUESTED. Failure to supply complete information will delay processing and may cause denial of the application.

PAPERWORK REDUCTION ACT NOTICE

This form is in accordance with the clearance requirements of Section 3507, PL 96-511, 12/11/80. The information you provide is used to establish that a transferee's receipt and possession of the firearm would be in conformance with Federal, State, and local law. The data is used as proof of lawful registration of a firearm to the manufacturer. The furnishing of this information is mandatory (26 USC 5822).

ATF FORM 1 (7560.1) (4-82)

BATF Form 1, page 2 of 4.

Form Approved: OMB No. 1512-0024 (8/31/8?)

DEPARTMENT OF THE TREASURY – BUREAU OF ALCOHOL, TOBACCO AND FIREARMS

APPLICATION TO MAKE AND REGISTER A FIREARM

(Submit in duplicate. See Instructions attached.)

TO: National Firearms Act Branch, Bureau of Alcohol, Tobacco and Firearms, Washington, DC 20226

The undersigned hereby makes application, as required by Sections 5821 and 5822 of the National Firearms Act, Title 26 U.S.C., Chapter 53, to make and register the firearm described below.

1. TYPE OF APPLICATION *(Check one)*

☐ a. TAX PAID. Submit with your application a check or money order for $200 made payable to the Department of the Treasury. Upon approval of the application, this office will acquire, affix, and cancel the required "National Firearms Act" stamp for you.

☐ b. TAX EXEMPT because firearm is being made on behalf of the United States, or any department, independent establishment, or agency thereof.

☐ c. TAX EXEMPT because firearm is being made by or on behalf of any State or possession of the United States, or any political subdivision thereof, or any official police organization of such a government entity engaged in criminal investigations.

2. APPLICATION IS MADE BY: ☐ INDIVIDUAL ☐ BUSINESS FIRM ☐ GOVERNMENT ENTITY

TRADE NAME *(If any)*

3a. APPLICANT'S NAME AND MAILING ADDRESS *(Type or print below and between the dots)*

b. IF P.O. BOX IS SHOWN ABOVE, STREET ADDRESS MUST BE GIVEN HERE

c. COUNTY

IMPORTANT: COMPLETE THE REVERSE SIDE. INDIVIDUALS (INCLUDING LICENSED COLLECTORS) MUST ALSO SUBMIT, IN DUPLICATE, FBI FORM FD-258, FINGERPRINT CARD.

d. TELEPHONE AREA CODE AND NUMBER

4. DESCRIPTION OF FIREARM (Complete items a through i)

a. NAME AND LOCATION OF ORIGINAL MANUFACTURER OF FIREARM (RECEIVER) *(If prototype, furnish plans and specifications) (See Instruction 2f.)*

b. TYPE OF FIREARM TO BE MADE *(Shortbarreled rifle, machine gun, destructive device, etc.)*

c. CALIBER, GAUGE OR SIZE *(Specify)*

d. MODEL

LENGTH *(inches)* e. OF BARREL: f. OVERALL:

g. SERIAL NUMBER *(See Instruction 2g.)*

h. ADDITIONAL DESCRIPTION *(Include all numbers and other identifying data which will appear on the firearm)*

i. STATE WHY YOU INTEND TO MAKE FIREARM *(Use additional sheet if necessary)*

5. APPLICANT'S FEDERAL FIREARMS LICENSE *(If any)* *(Give complete 15-digit number)*

First 6 digits	2 digits	2 digits	5 digits

6. SPECIAL (OCCUPATIONAL) TAX STATUS

a. ATF IDENTIFICATION NUMBER b. CLASS

7. EMPLOYER'S IDENTIFICATION NUMBER *(If applicable)*

IMPORTANT: GIVE FULL DETAILS ON SEPARATE SHEET FOR ALL "YES" ANSWERS IN ITEMS 8 AND 9

8. IS APPLICANT	YES	NO	9. HAS APPLICANT	YES	NO
a. Charged by information or under indictment in any court for a crime punishable by imprisonment for a term exceeding one year?			a. Been convicted in any court of a crime punishable by imprisonment for a term exceeding one year?		
b. A fugitive from justice?			b. Been discharged from the armed forces under dishonorable conditions?		
c. An alien who is illegally or unlawfully in the United States?			c. Been adjudicated as a mental defective or been committed to any mental institution?		
d. Under 21 years of age?					
e. An unlawful user of or addicted to marihuana or any depressant, stimulant or narcotic drug?			d. Renounced his or her citizenship, having been a citizen of the United States?		
f. A citizen of the United States					

UNDER PENALTIES OF PERJURY, I DECLARE that I have examined this application, including accompanying documents, and to the best of my knowledge and belief it is true, accurate and complete, and the making and possession of the firearm described above would not constitute a violation of Chapter 44, Title 18, U.S.C., Chapter 53, Title 26, U.S.C., Title VII of the Omnibus Crime Control and Safe Streets Act, as amended, or any provisions of State or local law.

10. SIGNATURE OF APPLICANT

11. NAME AND TITLE OF AUTHORIZED OFFICIAL OF FIRM OR CORPORATION *(If applicable)*

12. DATE

THE SPACE BELOW IS FOR THE USE OF THE BUREAU OF ALCOHOL, TOBACCO AND FIREARMS

BY AUTHORITY OF THE DIRECTOR, BUREAU OF ALCOHOL, TOBACCO AND FIREARMS, THIS APPLICATION HAS BEEN EXAMINED AND THE APPLICANT'S MAKING AND REGISTRATION OF THE FIREARM DESCRIBED ABOVE IS:

STAMP NUMBER

☐ APPROVED *(With the following conditions, if any)*

☐ DISAPPROVED *(For the following reasons)*

EXAMINER | DATE | AUTHORIZED ATF OFFICIAL | DATE

ATF FORM 1 (7560.1) (4-82)

BATF Form 1, page 3 of 4.

ADDITIONAL REQUIREMENTS

The Chief of Police, Sheriff, or other official acceptable to the Director must complete the "Law Enforcement Certification" below. If the applicant is an individual (including a licensed collector) a recent photograph must be attached in the space provided and FBI Form FD-258, Fingerprint Card, completed in duplicate, must be submitted.

1. PHOTOGRAPH

AFFIX
RECENT PHOTOGRAPH HERE
(Approximately 2" x 2")

2. LAW ENFORCEMENT CERTIFICATION (See IMPORTANT note below)

I certify that I am the chief law enforcement officer of the organization named below having jurisdiction in the area of residence of

____
(Name of Maker)

I have no information indicating that the maker will use the firearm or device described on this application for other than lawful purposes. I have no information that POSSESSION OF THE FIREARM DESCRIBED IN ITEM 4 ON THE FRONT OF THIS FORM WOULD PLACE THE MAKER IN VIOLATION OF STATE OR LOCAL LAW.

____
(Signature and Title of Chief Law Enforcement Officer—see IMPORTANT note below)

BY (See IMPORTANT note below) ________________________________
(Signature and Title of Delegated Person)

____
(Organization)

____
(Street Address)

________________________________ ____________
(City, State, and ZIP Code) (Date)

IMPORTANT: The chief law enforcement officer is considered to be the Chief of Police for the maker's city or town of residence, the Sheriff for the maker's county of residence; the Head of the State Police for the maker's State of residence; a State or local district attorney or prosecutor having jurisdiction in the maker's area of residence; or another person whose certification is acceptable to the Director, Bureau of Alcohol, Tobacco and Firearms, such as a U.S. Attorney. If someone has specific delegated authority to sign on behalf of the Chief of Police, Sheriff, etc., this fact must be noted by printing the Chief's, Sheriff's, or other authorized official's name and title, followed by the word "by" and the full signature and title of the delegated person.

GPO 890-681 ATF FORM 1 (7560.1)(4-82)

BATF Form 1, page 4 of 4.

Form Approved: OMB No. 1512 0027 (10/31/89)

DEPARTMENT OF THE TREASURY — BUREAU OF ALCOHOL, TOBACCO AND FIREARMS

APPLICATION FOR TAX PAID TRANSFER AND REGISTRATION OF FIREARM

DETACH THIS SHEET BEFORE COMPLETING FORM
INSTRUCTIONS

1. DEFINITIONS

a. FIREARM. The term "firearm" means: (1) a shotgun having a barrel or barrels of less than 18 inches in length; (2) a weapon made from a shotgun if such weapon as modified has an overall length of less than 26 inches or a barrel or barrels of less than 18 inches in length; (3) a rifle having a barrel or barrels of less than 16 inches in length; (4) a weapon made from a rifle if such weapon as modified has an overall length of less than 26 inches or a barrel or barrels of less than 16 inches in length; (5) any other weapon, as defined in b. below; (6) a machinegun; (7) a muffler or a silencer for any firearm whether or not such firearm is included within this definition; and (8) a destructive device. The term "firearm" shall not include an antique firearm or any device (other than a machinegun or destructive device) which, although designed as a weapon, the Director, Bureau of Alcohol, Tobacco and Firearms, or authorized delegate, finds by reason of the date of its manufacture, value, design and other characteristics is primarily a collector's item and is not likely to be used as a weapon.

b. ANY OTHER WEAPON. The term "any other weapon" means any weapon or device capable of being concealed on the person and from which a shot can be discharged through the energy of an explosive; a pistol or revolver having a barrel with a smooth bore designed or redesigned to fire a fixed shotgun shell; weapons with combination shotgun and rifle barrels 12 inches or more, less than 18 inches in length, from which only a single discharge can be made from either barrel without manual reloading; and shall include any such weapon which may be readily restored to fire. Such term shall not include a pistol or a revolver having a rifled bore, or rifled bores, or weapons designed, made, or intended to be fired from the shoulder and not capable of firing fixed ammunition.

c. TRANSFEROR. The term "transferor" means the registered owner of a firearm who is applying to transfer it.

d. TRANSFEREE. The term "transferee" means the person desiring to acquire the firearm.

e. PERSON. The term "person" means any individual, company, corporation, association, firm, partnership, joint stock company, trust or society other than a special (occupational) taxpayer qualified to deal in NFA firearms.

2. PREPARATION OF APPLICATION FOR TAX PAID TRANSFER AND REGISTRATION OF A FIREARM.

a. Reference §§5811 and 5812, National Firearms Act, United States Code, Chapter 53), persons seeking to transfer a firearm tax paid must complete, in duplicate, a separate application on this form for each firearm. The applicant transferor must furnish all the information called for on this application form.

b. If the transferor of a destructive device, machinegun, shortbarreled shotgun or short-barreled rifle is a licensee under 18 U.S.C., Chapter 44, and the transferee is anyone other than a qualified licensee under the National Firearms Act [special (occupational) taxpayer], the transferee must complete item 2 (Applicant Certification) on the reverse side of ATF Form 4 (5320.4). Item 3 (Law Enforcement Certification) must be completed for the transfer of any NFA firearm to an individual other than a special (occupational) taxpayer.

c. If the transferee is an individual, including a collector licensed under 18 U.S.C., Chapter 44, in addition to satisfying the requirements of 2b, above, a recent 2" x 2" photograph must be affixed in item 1 on the reverse of ATF Form 4 (5320.4) and completed FBI Form FD-258, Fingerprint Card, must be submitted in duplicate. The fingerprints should be taken by a person qualified to do so, and must be clear, unsmudged and classifiable.

d. All signatures required on this form must be entered in ink on both copies. Photocopies or other facsimile signatures are not acceptable. It is preferred that the form be prepared by the use of a typewriter, using carbon paper to make an exact duplicate. Pen and ink may be used, but under no circumstances will a form filled in by use of a lead pencil be accepted. All changes made on this form must be initialed and dated.

e. If the transferee is acquiring the firearm as other than an individual, the trade name and business address should be shown in item 2 on the face of ATF Form 4 (5320.4). Firearms may not be acquired as a part of the business inventory of a firearms licensee who is not a special (occupational) taxpayer. In such case, the home address of the transferee should be shown.

f. If the firearm is being transferred from an estate, item 3a. should show the transferor as: Executor's name, title (Executor (or executrix, administrator, administratrix) of the Estate of (name)), and the executor's address. Item 3c. should reflect the last address of the decedent and date of death.

g. If any questions arise concerning the preparation of this form, please contact the NFA Branch, Bureau of Alcohol, Tobacco and Firearms, Washington, DC 20226 or the nearest Alcohol, Tobacco and Firearms office.

3. DISPOSITION OF APPLICATION FOR TRANSFER AND REGISTRATION OF FIREARM. The transferor will forward both copies of the form to the National Firearms Act Branch, Bureau of Alcohol, Tobacco and Firearms, P.O. Box 73201 Chicago, IL 60673-7201. This office will deposit the tax and forward the application to the Bureau of Alcohol, Tobacco and Firearms, National Firearms Act Branch, Washington, DC 20226 for processing. If approved, the original of the approved form will be returned to the transferor for delivery to the transferee with the firearm, and this office will retain the duplicate. Approval of this form will effect registration of the firearm to the transferee. The transferor shall not under any circumstances deliver the firearm to the transferee until the approved form is received. This form must be retained by the transferee and be available for inspection by Government officers until such time as the firearm may later be transferred after approval by this office. (If the application is disapproved, the original with any accompanying check or money order, will be returned to the transferor with the reason for disapproval.)

4. APPROVED TRANSFER NOT PHYSICALLY ACCOMPLISHED. The transferor should mark the front of the form "VOID," sign and date the form and return the voided form to the Director. This will assure that the firearm will remain registered in the name of the transferor who submitted the application. Upon receipt of the voided form, instructions for obtaining a tax refund will be provided.

5. LATER TRANSFERS OF THE FIREARM. If the firearm is to be transferred later by the new owner, a new application form covering the proposed transfer must be filed with the Director.

ATF Form 4 (5320.4) (7-88)

This is the BATF Form 4, which covers transfer of an existing silencer between two parties. It must be completed in duplicate and submitted to the BATF along with two fingerprint cards and a $200 check. (Courtesy of BATF)

INSTRUCTIONS (Continued)

6. RATE OF TAX ON THE TRANSFER OF A FIREARM. The tax on the transfer of a firearm is $200, except that the transfer tax is $5 on any firearm classified as "any other weapon" as defined in instruction 1b.

7. PERSONS PROHIBITED FROM RECEIVING FIREARMS. Under 18 U.S.C., Chapter 44 and Title VII of Public Law 90-351, as amended, (18 U.S.C., Appendix), the following persons are prohibited from receiving a firearm, including an unserviceable firearm which has a frame or receiver: (1) fugitives from justice (any crime); (2) persons under indictment for, or who have been convicted of, a crime punishable by imprisonment, for a term exceeding 1 year; (3) narcotic addicts or unlawful drug users; (4) persons adjudicated mental defectives or mentally incompetent, or who have been committed to any mental institution; (5) veterans discharged under dishonorable conditions; (6) persons who have renounced their United States citizenship; (7) aliens illegally or unlawfully in the United States; and (8) where the transferor is a firearms licensee, persons under 21 years of age in the case of any firearm other than a shotgun or a rifle and under 18 years of age in the case of a shotgun or a rifle. In addition, section 5812 of the National Firearms Act requires that an application to transfer a firearm be denied if the transfer, receipt or possession of the firearm would place the transferee in violation of law. The term "law" in this statute includes Federal laws as well as State statutes and local ordinances applicable to the locality where the transferee resides.

8. INQUIRIES ABOUT STATUS OF A TRANSFER APPLICATION. The transfer application form will be reviewed and returned to the transferor promptly if additional information or corrections are required. However, because of the necessity for an FBI record check on an individual transferee's fingerprints, approximately 60 days must be allowed for processing such applications. Under the provisions of Section 6103(a), Title 26, United States Code, disclosure of any "return" or "return information" is generally prohibited to anyone except the person filing the return. Therefore, information about the status of a transfer application may be given only to the transferor. Such information cannot be divulged to the transferee.

9. PENALTIES. Any person who violates or fails to comply with any of the requirements of the National Firearms Act shall, upon conviction, be fined not more than $10,000 or be imprisoned for not more than 10 years, or both, in the discretion of the court. Moreover, any firearm involved in any violation of the provisions of the National Firearms Act or any regulations issued thereunder shall be subject to seizure and forfeiture. It is unlawful for any person to make or cause the making of a false entry on any application or record required by the National Firearms Act knowing such entry to be false.

NOTE: All provisions of Title I of the Gun Control Act must also be complied with, including the recordkeeping requirements for licenses and retention of ATF Form 4473 (5300.9) by licensees for dispositions to nonlicensees.

SPECIAL NOTE: A direct interstate transfer will be approved only to a licensee who is also a special (occupational) taxpayer qualified to deal in the particular type NFA firearm; or a government entity; or a lawful heir; or, in the case of a firearm which has been designated as a "curio or relic," to a licensed collector.

PAPERWORK REDUCTION ACT NOTICE

This information request is in accordance with the Paperwork Reduction Act of 1980. The information you provide is used in applying to transfer serviceable firearms taxpaid to persons other than special taxpayers qualified to deal in NFA firearms or government entities. Data is used to identify transferor, transferee, and firearm, and to ensure legality for transfer under Federal, State and local laws. The furnishing of this information is mandatory. (26 U.S.C. 5812)

The following information is provided pursuant to Section 3 of the Privacy Act of 1974 (5 U.S.C. §552(e)(3)):

1. *AUTHORITY.* Solicitation of this information is made pursuant to the National Firearms Act (26 U.S.C. §5812). Disclosure of this information by the applicant is mandatory for transfer of an NFA firearm, unless the transfer is otherwise exempt from tax.
2. *PURPOSE.* To insure payment of the tax imposed by 26 U.S.C. §5811; to insure that the transfer would not be in violation of law; and to effect registration of the firearm.
3. *ROUTINE USES.* The information will be used by ATF to make the determinations set forth in paragraph 2. In addition, to effect registration of the firearm, information as to the identification of the firearm, date of registration, and the identification and address of person entitled to possess the firearm will be entered into the National Firearms Registration and Transfer Record. No information obtained from an application, registration, or records required to be submitted by a natural person in order to comply with any provision of the National Firearms Act or regulations issued thereunder, shall, except in connection with prosecution or other action for furnishing false information, be used, directly or indirectly, as evidence against that person in any criminal proceeding with respect to a violation of law occurring prior to or concurrently with the filing of the application. The information from this application may only be disclosed to Federal authorities for purposes of prosecution for violation of the National Firearms Act.
4. *EFFECTS OF NOT SUPPLYING INFORMATION REQUESTED.* Failure to supply complete information will delay processing and may cause denial of the application.

*U.S. Government Printing Office: 1988—202-929/86726 ATF Form 4 (5320.4) (7-88)

BATF Form 4, page 2 of 4.

Form Approved: OMB No. 1512-0027 (10/31/89)

DEPARTMENT OF THE TREASURY – BUREAU OF ALCOHOL, TOBACCO AND FIREARMS

APPLICATION FOR TAX PAID TRANSFER AND REGISTRATION OF FIREARM

SEE INSTRUCTIONS ATTACHED.
TO BE SUBMITTED IN DUPLICATE
TO: National Firearms Act Branch
Bureau of Alcohol, Tobacco and Firearms
P.O. Box 73201
Chicago, IL 60673-7201

2a. TRANSFEREE'S NAME AND ADDRESS (If transferee is a Special (Occupational) Taxpayer who is acquiring firearm for personal use, rather than as part of his business inventory, show personal name below and check here: ☐)

1. TYPE OF TRANSFER *(Check one)* *(See instructions 1 and 6)*
☐ $5 ☐ $200

2b. TRADE NAME *(See instruction 2e)*

2c. COUNTY

3a. TRANSFEROR'S NAME AND MAILING ADDRESS (If the firearm is registered under your trade name, enter your trade name. EXECUTORS: See instruction 2f.)

Submit with your application a check or money order for the appropriate amount made payable to the Department of the Treasury. Upon approval of this application, this office will acquire, affix and cancel the required "National Firearms Act" stamp for you. *(See Instruction 6)*

3b. NUMBER, STREET, CITY, STATE AND ZIP CODE IF DIFFERENT FROM ITEM 3a.

3c. IF APPLICABLE: DECEDENT'S NAME, ADDRESS, AND DATE OF DEATH

The above-named and undersigned tranferor hereby makes application as required by Section 5812 of the National Firearms Act to transfer and register the firearm described below to the transferee.

4. DESCRIPTION OF FIREARM *(Complete items a through h)*

a. NAME AND ADDRESS OF MANUFACTURER AND/OR IMPORTER OF FIREARM	b. TYPE OF FIREARM *(Short-barreled rifle, machine gun, destructive device, any other weapon, etc.)*	c. CALIBER, GAUGE OR SIZE *(Specify)*	d. MODEL
			LENGTH *(Inches)* e. OF BARREL: f. OVERALL:
			g. SERIAL NUMBER

h. ADDITIONAL DESCRIPTION OR DATA APPEARING ON FIREARM *(Attach additional sheet if necessary)*

5. TRANSFEREE'S FEDERAL FIREARMS LICENSE *(If any)* *(Give complete 15-digit number)*

First 6 digits	2 digits	2 digits	5 digits

6. TRANSFEREE'S SPECIAL (OCCUPATIONAL) TAX STATUS

a. ATF IDENTIFICATION NUMBER	b. CLASS

7. TRANSFEROR'S FEDERAL FIREARMS LICENSE *(If any)* *(Give complete 15-digit number)*

First 6 digits	2 digits	2 digits	5 digits

8. TRANSFEROR'S SPECIAL (OCCUPATIONAL) TAX STATUS

a. ATF IDENTIFICATION NUMBER	b. CLASS

UNDER PENALTIES OF PERJURY, I DECLARE that I have examined this application, and to the best of my knowledge and belief it is true, correct and complete, and that the transfer of the described firearm to the transferee and receipt and possession of it by the transferee are not prohibited by the provisions of Chapter 44, Title 18, United States Code; Chapter 53, Title 26, United States Code; or Title VII of the Omnibus Crime Control and Safe Streets Act, as amended; or any provisions of State or local law.

9. SIGNATURE OF TRANSFEROR *(Or authorized official)*	10. NAME AND TITLE OF AUTHORIZED OFFICIAL *(Print or type)*	11. DATE

THE SPACE BELOW IS FOR THE USE OF THE BUREAU OF ALCOHOL, TOBACCO AND FIREARMS

BY AUTHORITY OF THE DIRECTOR, THIS APPLICATION HAS BEEN EXAMINED, AND THE TRANSFER AND REGISTRATION OF THE FIREARM DESCRIBED HEREIN AND THE INTERSTATE MOVEMENT OF THAT FIREARM, WHEN APPLICABLE, TO THE TRANSFEREE ARE:

STAMP NUMBER

☐ APPROVED *(With the following conditions, if any)*

☐ DISAPPROVED *(For the following reasons)*

SIGNATURE OF DIRECTOR, BUREAU OF ALCOHOL, TOBACCO AND FIREARMS	DATE

ATF Form 4 (5320.4) (7-88)

BATF Form 4, page 3 of 4.

CERTIFICATIONS

If the transferor of a destructive device, machinegun, short-barreled shotgun or short-barreled rifle is a Federal firearms licensee, and the transferee is anyone other than a licensee qualified to deal in the firearm to be transferred, the transferee must sign the Applicant Certification (item 2 below) in the presence of the law enforcement officer signing item 3 below. The Law Enforcement Certification (item 3 below) must be completed for the transfer of any registered firearm to an individual other than a licensee qualified to deal in the firearm to be transferred. In addition, the individual transferee must affix a recent photograph (taken within the past year) in item 1 and submit, in duplicate (to the transferor) two completed copies of FBI Form FD-258, Fingerprint Card. (See Important note below.)

1. PHOTOGRAPH

AFFIX
RECENT PHOTOGRAPH HERE
(Approximately 2" x 2")

2. APPLICANT CERTIFICATION

I, ______________________ (Name of Transferee), have a reasonable necessity to possess the device or weapon described on this application for the following reason(s) ______________________

____

and my possession of the device or weapon would be consistent with public safety (18 U.S.C. 922(b) (4) and 27 CFR 178.98).

UNDER PENALTIES OF PERJURY, I declare that I have examined this application, and to the best of my knowledge and belief it is true, correct and complete, and that receipt and possession of the firearm described on this form will not place me in violation of the provisions of Chapter 44, Title 18, U.S.C.; Chapter 53, Title 26, U.S.C.; or Title VII of the Omnibus Crime Control and Safe Streets Act, as amended, or any provisions of State or local law.

______________________ (Signature of Transferee or official authorized to sign for firm) ______________ Date

3. LAW ENFORCEMENT CERTIFICATION (See IMPORTANT note below)

I certify that I am the chief law enforcement officer of the organization named below having jurisdiction in the area of residence of ______________________ (Name of Transferee). I have no information indicating that the transferee will use the firearm or device described on this application for other than lawful purposes. I have no information that the receipt and/or possession of the firearm described in item 4 of this form would place the transferee in violation of State or local law.

______________________ (Signature and Title of Chief Law Enforcement Officer — See IMPORTANT note below)

______________________ ______________ (Date)

______________________ (Organization and Street Address)

______________________ (County) ______________ (Phone — include area code)

IMPORTANT: The chief law enforcement officer is considered to be the Chief of Police for the transferee's city or town of residence, the Sheriff for the transferee's county of residence; the Head of the State Police for the transferee's State of residence; a State or local district attorney or prosecutor having jurisdiction in the transferee's area of residence; or another person whose certification is acceptable to the Director, Bureau of Alcohol, Tobacco and Firearms. If someone has specific delegated authority to sign on behalf of the Chief of Police, Sheriff, etc., this fact must be noted by printing the Chief's, Sheriff's, or other authorized official's name and title, followed by the word "by" and the full signature and title of the delegated person.

ATF Form 4 (5320.4) (7-88)

BATF Form 4, page 4 of 4.

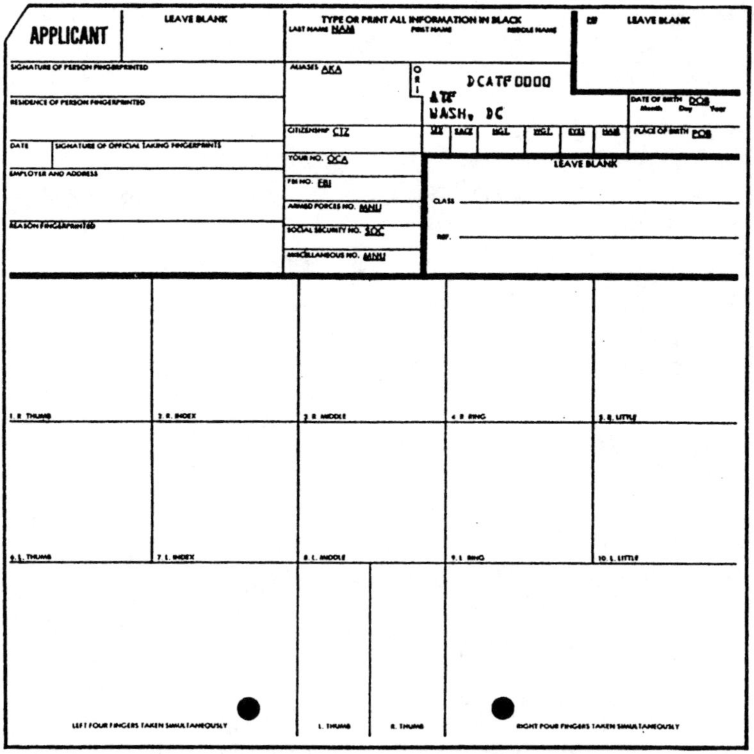

APPLICANT

LEAVE BLANK

TYPE OR PRINT ALL INFORMATION IN BLACK

LAST NAME NAM FIRST NAME MIDDLE NAME

FBI LEAVE BLANK

SIGNATURE OF PERSON FINGERPRINTED

ALIASES AKA

ORI DCATF0000 ATF WASH, DC

RESIDENCE OF PERSON FINGERPRINTED

DATE OF BIRTH DOB Month Day Year

DATE SIGNATURE OF OFFICIAL TAKING FINGERPRINTS

CITIZENSHIP CTZ

SEX RACE HGT. WGT. EYES HAIR PLACE OF BIRTH POB

EMPLOYER AND ADDRESS

YOUR NO. OCA

LEAVE BLANK

FBI NO. FBI

CLASS

ARMED FORCES NO. MNU

REASON FINGERPRINTED

SOCIAL SECURITY NO. SOC

REF.

MISCELLANEOUS NO. MNU

1. R. THUMB 2. R. INDEX 3. R. MIDDLE 4. R. RING 5. R. LITTLE

6. L. THUMB 7. L. INDEX 8. L. MIDDLE 9. L. RING 10. L. LITTLE

LEFT FOUR FINGERS TAKEN SIMULTANEOUSLY L. THUMB R. THUMB RIGHT FOUR FINGERS TAKEN SIMULTANEOUSLY

To make or transfer a silencer, you must file two completed and signed copies of this FBI Form 258, the universal fingerprint card, with your completed application. (Courtesy of FBI)

BATF FORM 1

This is the form you file in duplicate with the feds before you actually build a silencer. It will register your silencer with the federal government. Filing this form requires your photograph, fingerprints, a local law-enforcement sign-off, and payment of a $500 manufacturing tax.

It is very important that you do not start any construction on your silencer until this form comes back approved by the appropriate Bureau of Alcohol, Tobacco and Firearms (BATF) official. Under current law, the BATF considers silencer parts to be controlled items under the National Firearms Act. The point is, if you start assembling these NFA parts prior to having that approved Form 1 in your possession, the feds could claim prima facie evidence of intent to manufacture an unlicensed silencer.

If that sounds stupid, petty, nasty, silly, and legally serious, that's because it is. The only nonsilly part of it is when they fine you or throw your ass in jail. As I've written and said several hundred times, most federales have no sense of humor and take themselves far too seriously.

BATF FORM 4

You use this form when you buy a legally registered silencer from a dealer or manufacturer. They usually will provide you with this form. Filing it transfers registration and ownership of the silencer from the previous owner to you. You must include your photograph, fingerprints, a local law-enforcement signature, and payment of a $200 transfer tax.

FBI FORM 258

This is the universal fingerprint card that your local or state police personnel will use to record your fingerprints for the BATF files in Washington. Supposedly, each applicant's prints are checked, but I know a retired FBI employee who told me that it is rarely done. "Not nearly enough time or people or, frankly, concern" is how he put it to me.

Those are the forms. Now, here is the proper drill that you should use if you want to be all tickety-boo legal and have that homemade silencer registered with the BATF. Before building the silencer, fill out Form 1 after you get your picture and fingerprints. Get the endorsement signature of your favorite local law-enforcement official. The feds like chiefs of police, county sheriffs, chief prosecutors, or their designated minions to sign for this. Obviously, ask someone you know and who knows you; it's easier that way.

If they ask you what in hell you want a silencer for, remember that they are *not* there to give you permission to own the silencer. Some local official's personal prejudices about gun and silencer control is not necessarily the law. You are going through a legal process here, not getting permission. What the local officer is signing and attesting to is that you are not of bad character and that he has no information that would lead him to believe you would use the silencer illegally. He is also attesting that your ownership of the silencer doesn't violate any local laws.

This official does not incur any legal liability by signing this form for you. On the other hand, no law requires him to sign it, either. So as I say, make friends with someone in the upper echelon of your town or county police department. Remember, this official must have law enforcement jurisdiction within your area of legal residence.

Next, write a check for $200 to the Department of Treasury to pay your tax. Fill out the rest of the form and sign it. Make at least one photocopy of the completed paperwork, then mail the two originals to the address indicated on the form by certified mail.

If you're buying someone else's already registered silencer, you go through the exact same procedure except you use a BATF Form 4. Everything else is the same. In either case, you will probably hear from BATF in two or three months.

Silencer fans, that's all you need to do to have your very own legally registered silencer.

If you're still not comfortable with all of this, or you just have a few more questions, my advice is to find your nearest user-friendly Class III dealer or Class II manufacturer. If you need some help in this area, please see the "Sources' Apprentice" chapter of this book, or write to me for a free referral to a good, honest, reliable, trustworthy, and friendly Class III dealer or Class II manufacturer in your area.

Remember, you must abide by all local and state laws, obtain a BATF Form 1, and pay all of the required taxes before you even think about touching a tool to a component. If you don't and are caught, your buns are deep grass and will be easy mowing for the feds. It's only former presidents and prevaricating politicians who can get away with "I forgot . . . I don't remember . . . I didn't know that was wrong."

You will go to jail.

I was talking with the senior partners of the law firm of Rimkus, Brooks, Simpson, Montfort, Neal, and Dilligaf during a legal seminar. We were discussing the topic of Constitutional protection and the Second Amendment at Smuggler's Inn Lounge when the topic of politicians came up. "Remember, George, most politicians are simply pimps who went to law school," reminded partner Vic Dilligaf. "Many of those pimps also end up on the wrong side of the criminal justice system—as in being in charge of it."

Most of the criminal justice system is made up of their rules. Those rules often disregard our Constitution, especially the Second Amendment. And it is easier for them to convict people by violating our Constitution than by complying with it. Sometimes they can't convict you without violating our Constitution.

Remember, rules and constitutions are two different things. So be sure that you follow their rules, because you can be damned sure that many of them aren't following our Constitution.

Follow my instructions as well as the letter of the law and you'll be fine, safe, and happy with your new homemade silencer. Speaking of which, we've put this off long enough. Let's turn some pages and build some silencers.

Next, what you'll need, how to use them, and how to find them . . . tools, components, and sources.

2 No Tool Like an Old Tool

Despite run-amok technology, Tim "The Tool Man" Taylor assures me that there are no newbies in the home silencer maker's kit. The same hardware listed in *The Hayduke Silencer Book: Quick and Dirty Homemade Silencers* will suffice to cut, bend, stuff, fasten, and otherwise fabricate your very own homemade silencer design or one from this book.

This is not to endorse the idea that you can't look. Keeping an open eye and mind to the changing tool technology is just good smarts. Always stay current with technical improvement. To do otherwise would be foolish. As Forrest Gump's wonderfully pragmatic mother noted, "Stupid is as stupid does," or something as profound as current psycho-sociology babble gets in these days of dumb-down diversity.

TOOLS

My advice is to peruse the catalogs, the home-builder stores, and other sources for tools. There are always new ones out there, and only a fool passes on technology. As a young man I recall hearing the wise words of the ever-exotic Lola Montez, confidant of Indiana Jones and a famed adventuress in her own realm, who told me, "George, there's no tool like an old tool."

You aren't going to need a workshop full of tools to build

these silencers. You probably have most, if not all, of them in your home now. Here is my list of the common tools you'll need:

1. A small electric hand drill, cordless or not, and drill bits. My advice is to buy one of those plastic-cased assortment packs of bits.
2. A good hacksaw. Actually, a hacksaw is better than a small power saw because you won't be doing much volume or heavy-duty cutting. The more precise hand model is for you.
3. A good, solid ball-peen hammer is essential, as is a good rubber mallet that won't scratch or mar the fine metal finish on your gun.
4. Wrenches. I suggest both a small- and medium-sized crescent wrench. Also, you could probably make good use of a plumber's slip-joint wrench, and everyone's old standby, a good set of vice grips.
5. A rivet gun and rivets will be useful for some designs. You'll be working with light fastening, so nothing fancy is required.
6. A set of files. Again, the usual prepackaged selection will be fine.
7. A ruler, a metal marking punch, and scissors round out the list.

MATERIALS

The list of materials you'll need is pretty basic, too. Since I have no idea what type or size silencer you're going to build, I will leave the sizes, amounts, and quantities that you purchase up to you. The plans in various design chapters tell you the specific items you'll need for each design, so consult them before you shop.

What I am offering here are basic, generic, and common materials. Like the tools, these are readily available in hardware, automotive, and home-improvement shops. Common caution suggests that you spread your purchases over several stores to avoid suspicion. And if I were you, I would not discuss the purpose for your purchases with anyone.

You will need six rolls of black electrician's tape and a roll of duct tape. Get good-quality tape, please. Pass up the bargain cheapie stuff—it's generally not tough, strong, or very sticky. Epoxy is very useful to have as well.

You'll also need lengths of both PVC and aluminum pipe, in smaller sizes, e.g., 3/4-, 1-, and 2-inch. You can make the determinations based on what size silencer you want to build for what caliber gun. The plans in each specific chapter will guide you in this.

Buy either aluminum or copper window screening. It has to be true metal, so don't buy plastic screening. Again, the amount will depend on which plans you choose.

Buy some Chore Boy pure copper cleaning sponges, too. They make great internal packing for silencers.

A selection of the appropriately sized frieze plugs, metal washers, rubber stoppers, and the like will be useful in most of the designs, as you'll soon read.

3 Technical Things You Need to Know

Clandestine paramilitary operations have given silencers a very sinister image that has been mirrored by our hypocritical entertainment/news media, violent society's pimpmeister. Stripped of personification and man-directed usage, a silencer is simply hardware put to a purpose. A silencer is a sound muffler just like the one on your car's engine.

Until the invention of gunpowder, the hardware and mechanics of both hunting and war were relatively quiet. Victims of war and hunting, human and otherwise, created most of the environmental noise pollution until some Chinese experimenters favored us with gunpowder long before even George Burns was born. The next step was to quiet the explosive noise of gunpowder, a task which attracted no end of experimenters, tinkerers, and engineers. Pioneers include Reade M. Washington and Alfred Capy, who were issued a patent in 1900 for a "muzzle attachment to muffle gunfire." Joseph E. Bissell designed and patented a basic silencer in 1902, while prolific gun designer Samuel McClean turned out four different patented designs in 1905.

The man generally credited with designing the first commercially successful and practical silencer was Hiram Percy Maxim in 1908. Since then, a plethora of designers has marched in and out of the patent office. There is a wealth of historical silencer design technology, by the way, in Don

RELATIVE SOUND LEVELS*

Noise Level 3-6' from Source	Effect	DB Level	Weapon Noise Level
		140	
Air Raid Siren		135	
Pneumatic Hammer	Painful	130	
Thunder		125	
Chain Saw		120	
Rock Concert	Uncomfortable	115	
Gas Mower		110	
Router		105	
Loud Classical Music		100	.45 and .380 ACP, unsilenced
Newspaper Press		95	
Food Blender		90	.22 LR, unsilenced
Loud Singing	Ear Damage	85	
Electric Typewriter		80	
Normal Conversation		75	.45 and .380 ACP Sionac, silenced
Office	Annoying	70	.22 LR silenced
		65	
Background Music		60	
		55	
Average Home		50	
		45	
		40	
		35	
Country Home		30	
		25	
Whisper		20	
		15	
Light Breeze		10	
Threshold of Hearing		5	

*Levels are approximate

This comparative chart shows relative sound levels from a variety of sources for comparison to silencer-equipped firearms. Credit: Manculich Research Associates.

Thomas' rare and wonderful books on silencer patents, e.g., full descriptions, sketches, etc. Although his books are now out of print, many engineering, technical, and military libraries have copies. Check with the interlibrary loan department of your local library for details on obtaining this valuable design resource.

Today there is a small group of silencer builders who are truly professional in their methodology and approach. Their silencers are marvels of design and manufacturing technology. They are also expensive. They have to be because the retail price of each of their units includes a share of the R&D costs.

The price of this book is your sole R&D cost. Following are 10 designs that I have chosen for you after consulting with a group of today's real silencer experts, including some of the very same folks I alluded to in the previous paragraph.

These designs range from the quick-and-dirty simple of prison improvisation to the more difficult design requiring home workshop tools, a more-than-room-temperature IQ, and an hour or two of your time.

All of these designs can be built in your own home workshop or even at the kitchen table, just as I described in my first silencer book. In fact, dig out your copy of the first book when you go shopping for tools, ingredients, and equipment, because it's all the same stuff as before.

What? You loaned your copy of my first silencer book to your Aunt Nancy and she hasn't returned it yet? In the interest of family peace, and not just to fatten this book with reprint, I've included a condensed version of what you'll need (see box, next page).

Once you've checked my shopping list, you might be tempted to ask why not simply buy a silencer kit? Some intrepid companies still bravely advertise them in the appropriate periodicals. Recall my words of caution from the last book, i.e., our fearless feds sometimes go on witch-hunts using easily available UPS records that trace these kits and parts from shipper to your home.

Our brave New World Order Police also purloin phone,

Professionally produced silencers are usually gorgeously machined and gorgeously expensive. Many of them are so complicated that they baffle even the experts.

Nonsense. Your silencer will be as inexpensive and simple as you wish it to be. There are some common items you will need, though, to construct the silencers in this book.

Ammunition. You already know that most ammunition is supersonic, thus adding the sonic boom effect to your problem of quieting gunfire. You need to bring that round down to subsonic levels. This is done either by using an already subsonic round, like the .45 ACP, or by porting your gun's barrel into the silencer to bleed off gasses, thus slowing a supersonic round to subsonic. There is another very easy way, which I recommend. Simply buy standard or subsonic velocity .22 or 9mm ammunition in the first place. Caution: Do not tell them why you want this type of ammo. Some gun shop denizens suck up to BATF thugs by ratting on nice folks like you.

Sights. You may have to remove the front sight from your rifle or pistol to accommodate the silencer tube. Even the original sights on many commercial cans are useless because of the silencer's bulk, so they are removed to facilitate engineering.

It is likely that your silencer will be concentric, meaning it will render the front sight useless. Therefore, you are either going to have to reinstall a front sight on the silencer and realign your sights, or mount a telescope sight on the weapon.

computer, and other records illegally from both sellers and buyers. It's a nasty, tangled, evil web that our masters weave to ensnare the unwary tinkerer who innocently buys silencer parts with no evil intent.

Hogwash, you say? Forget Pearl Harbor . . . remember Waco!

If you're going to be tinkering with silencer kits, I suggest that you do so quietly, and, other than the federal legalities, the fewer folks who know, the better it will be for you.

There is another danger in using these do-it-yourself units. Some of the parts kit technology is even more dated than the historical designs in Don Thomas' books. Other kit designs are just plain scary, more dangerous to the user than the target. Often, they are poorly made of inferior materials, with no alignment control, i.e., just not safe at all.

But if you're going to buy kit parts, I have three suggestions. First, be sure you're doing so legally. Then, be sure you're buying safe, modern-design, quality parts from a competent seller. Finally, be sure of suggestions one and two a lot more.

Remember, too, when you're working with most of the materials mentioned in this book, that they are generally home-workshop-grade materials, not heat-treated steel, aircraft-grade aluminum or armor plate. You will bend or break aluminum or thin steel if you apply too much muscle or torque.

With all that heavy legal/safe effluvia out of the way, a few more technical points will help you understand what you're about to do. The major job of a silencer, or suppressor, is to regulate and restrict the flow of noisy explosive gasses caused by the explosion of a cartridge. This is achieved usually through multichambered and baffled silencer designs whose genesis is engineering and physics. Not knowing much about either discipline, I'm not going to attempt to explain that technical facet. Instead, this book will show you workable designs you can build yourself. If you want to learn about the physics, engineering, interior ballistics, and other technical details, it's all available. See the bibliography on page 83.

Don't expect to build a totally quiet firearm silencer. Even the big boys and the technowonks don't achieve that, because there is no such thing as a totally quiet silencer. The sounds you don't hear guns make on your home or cineplex screen are the result of soundtrack magic, not silencer technology.

In fact, the real silencer designers call their products "sound suppressors," as that is a far more accurate name for them. Their jargon name for them, by the way, is "cans." I will continue to call the units silencers because that's what old Mr. Maxim called them.

A silencer must allow the bullet to exit the firearm's barrel and the attached silencing unit with ballistic integrity (a fancy term meaning the round doesn't touch the sides of the can on the way out) so that accuracy is maintained. The silencer must also delay and smooth the passage of the hot gases that follow the bullet to avoid the BANG explosion that occurs when the gases exit the barrel.

As a humane little quality-control tip, let me suggest that when you design and build your silencer, be certain to allow an adequate opening for the bullet's passage. The alternative is to plug the exit area, which creates a very dangerous buildup of pressure. This pressure will eventually explode, and it usually follows Tugmutton's Law: *The damn thing will blow up in your stupid face or the bolt will come roaring back through your eye.* The point of Tugmutton's creed is to impress upon you to *be careful*. Speaking of care, that famed woodsman and rustic philosopher, Terry Sink, pointed out, "There're two ways to shoot yourself in the foot . . . metaphorically and physically. And, despite what you hear, one hurts a helluva lot more than the other."

With that safety so blessed, you need to decide which of your guns you wish to equip with a silencer—rifle or handgun. Generally, silencers are more effective on rifles than on pistols of the same caliber. That's because the longer barrel distance allows the gasses to smooth out and dissipate more slowly and for the powder to burn more evenly, hence, less BOOM and more fisssss . . .

Obviously, a silencer for a pistol is far more efficient than one for a revolver, although some designers have built special housings to enclose the cylinder of a revolver and create a fairly quiet unit.

The experts discourage trying to put silencers on truly

high-velocity rifles, e.g., .223, .308, etc. To properly suppress such a muzzle blast and powerful high-velocity round requires a very large silencer unit and a great deal of barrel porting to slow down the round to near subsonic travel. However, some of the experts do recommend a .44 Magnum rifle as somewhat ideal for a silencer. Otherwise, stick to the "low and slow" rule that I explained in my earlier book.

As caliber is a basic consideration, your choices will be within the .22, .25, .32, .380, 9mm, and .45 calibers for handguns. Your long gun choices are far more numerous, but some of the more popular are .22, .223, 7.62mm, .30/06, and .308. You'll note that all of those popular calibers are of the high-velocity persuasion, and you will have to port the barrel to bleed speed. Keep that in mind. Whatever your situation or choice, remember those early-on basics I explained, in that some calibers are far more suited to silencers than others. For example, the .45 ACP round is ideal because it is already subsonic. On the other hand, the .223 round is supersonic and therefore less adaptable for the silencer mode.

The most conclusive statement I can make about caliber choice and silencers is that gravity always wins.

The general construction rules given in the first book apply here. Assuming you have that basic knowledge, I will include specific instructions dedicated to each of the specific designs that are only a few more pages away. Now that you have the temptation, the legalities, the tools, the tactics, and the technicalities, let's consider sources before we tackle those titans of homemade silencer design in the how-to chapters.

4
The Sources' Apprentice

Walt Disney's cold-shoulder legal condition not withstanding, I have a creepy hope that I don't get litigationed for spilling a titular bucketful of bad pun. Can we sweep that aside . . . quietly?

The folks and companies listed below sell useful equipment, parts, supplies, and information if you are going to make a legal silencer for your own use. My listing them does not constitute a personal endorsement. I have been somewhat selective, i.e., there are some notable rip-off artists whom I have not listed. Be aware, though, that companies do come and go, but that all addresses were current when this book was written.

All of these folks are legal, within the system, yet discreet about your inquiry. Some maintain no sales records or mailing lists, for whatever reason I cannot imagine. Feel secure to contact them within these contexts.

GUN LIST
700 E. State Street
Iola, WI 54990

GV METAL PRODUCTS
388 Government Street
Box 842
Burns Lake, BC
Canada VOJ 1EO

MAC, INC.
P.O. Box 597
Copperhill, TN 37317

MACHINE GUN NEWS
P.O. Box 759
Hot Springs, AR 71902-0759

PALADIN PRESS
P.O. Box 1307
Boulder, CO 80306

RPB
P.O. Box 367
Avondale Estates, GA 30002

SHOOTERS EQUIPMENT COMPANY
P.O. Box 517
Richland, SC 29675

SHOTGUN NEWS
P.O. Box 669
Hastings, NE 68902-0669

SUPPRESS-ON
P.O. Box 09161B
Detroit, MI 48209

And now, gentle reader, it's time to give you the designs.

5 Pamela Pisspig's Porcine Peashooter Popper

When she was about to terminate her last marriage to Bob Ray Grabbass, the famed feminist/anarchist Pamela Pisspig reportedly told him, "There are no guarantees in a relationship. If that's what you want, a long-term guarantee, then, go shack up with a Sears Diehard." I wonder what Pamela thinks of Bruce Willis.

Thanks to a referral from Hassan Farouk's North American business agent, I know what Ms. Pisspig's expertise is when it comes to silencer design. In a word, impressive. Pamela calls her own design "The Motorcycle Brake Tube" silencer.

Her construction steps are as follows. Purchase a standard slip-on gun barrel collar with 1/2-inch x 20 threads from one of the many such ads in that anarchist liberation publication *Shotgun News*.

Next, go to a people-friendly and tight-lipped machine shop and purchase a 6 1/2-inch piece of 13/16-inch inside diameter (ID) steel tubing. Have one end threaded for a removable cap. They should drill a centered 1/16-inch hole in this cap. But you will drill this hole out further to 15/64 inch.

Purchase a 1/2-inch x 20 NF machine nut at another location. File down the edges evenly until it slips snugly into the 13/16-inch tube. Weld it in place yourself or have a competent friend do it for you. The unit must be centered, and the weld must be airtight, obviously.

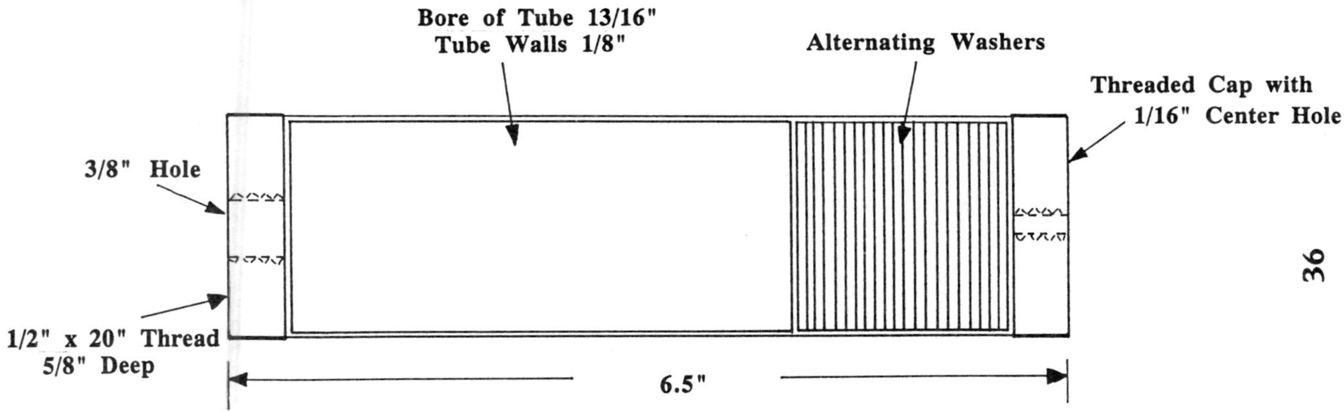

Figure 1. At last, an environmentally correct recyclement for previously owned motorcycle brake tubes. Like designer Pamela Pisspig, these units have a great body for silent, but deadly, discharge. Credit: Uncle Wally and Chiquita Bimbo.

Using medium-grade sandpaper, polish a 4-inch piece of standard 1/2-inch plastic water pipe until it will slip easily into the 13/16-inch tube. Slide it to the bottom of the tube. You now have a nice plastic bushing to support your column of washers that will form the expansion and gas dissipation chamber.

You build this by sliding a 5/16-inch flat washer down against that plastic pipe bushing. Next, straighten a 1/2-inch lock washer by using pliers to "take out" the lock. Slip that down over the first washer. Then, place another 1/2-inch flat washer on top of that. Now, alternately using a 1/2-inch flat and a 1/2-inch straightened lock washer, fill the tube in front of the plastic bushing.

The final step is to screw that threaded end cap with the 15/64-inch centered hole onto the threaded end of the tube, and, as Ms. Pisspig would say, "You're ready to do what you gotta do." Why did she call this a Motorcycle Brake Silencer? You'll have to ask her that. Does this design work? Yes. It was tested at the Enemies of Land Rapists firing range in Crushed Testicle, Utah. It produced a sound reduction of 30 dB when tested by Dr. Adolf Goering, a former BATF inspector, under field conditions.

Meantime, a friend of mine tested a registered homemade one just like this design and cut the sound of his gunfire by 28 dB.

6

Cuban Juan's Caribe Calladero

Reverend Gene met this Cubano dude in the joint, where all things large and small, legal and illegal, lethal and not, are possible. That's why the good Reverend knew you'd be interested in the quick-and-easy design that the Cuban con had built and used in some venue of incarceration over the years. In another chapter or two, I'll be showing you one of Rev Gene's own designs.

If he did it inside illegally, you can surely do it outside, and legally, as well. To replicate this design you need a lathe, a hand drill, and a soldering gun or small welding unit. You also need 5 inches of steam pipe, 6 inches of steel pipe about an inch more in diameter than the steam pipe, two pipe caps, and some steel wool.

First, you need to decide what gun this unit will be attached to and how, e.g., threading the barrel or using clamps, electrical tape, or compression fittings. The caliber of the firearm will determine the diameter of the hole you'll need to drill in the end cap to allow the passage of the bullet.

As this is a ported-barrel design, the next step is to drill 1/8-inch holes at about a 52-degree angle about 3/4 inch apart along the length of the steam pipe. You need to make four rows of them.

Then, wrap or pack steel wool around the ported pipe to a point that it will fit snugly inside the outer steel pipe. If you

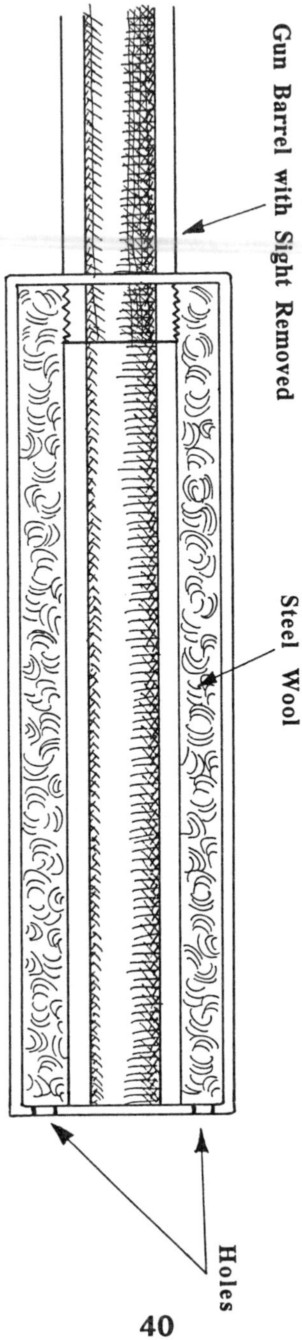

Figure 2. Cuban Juan's "Calladero de Caribe" (Caribbean silencer) is a masterfully modest and effective unit made quick 'n' dirty with easily available materials, tools, and methods. Credit: C.O. Jones, courtesy of Rev Gene.

don't like steel wool, you can substitute Chore Girl/Boy sponges at this step.

Drill a hole to match the caliber of the firearm into one end cap, centered, of course, then drill the other end cap to fit the end of the gun, or barrel threads, or compression unit, or whatever you've decided to use to attach the silencer. Be certain these holes are centered and aligned with the barrel of the gun, for obvious reasons. You can check alignment with a wood dowel before final assembly.

Pack it all together, mount it, and, as Cuban Juan would say, "¡Tiene usted un bien dia!" It is not important that your mark have one, however. After all, isn't that why they're called marks?

Your shot.

7 The Quirk Can

Thomas D. Quirk III amalgamated this design from others that he knew from his own background, plus the refinement of some from my first book. He refers to this as a hybrid, and I am happy to share the fruits of his design labor. Thus, I am pleased to present, for the first time in instructional design history, the Quirk Can.

His first step is to drill evenly spaced (1/2 to 3/4 inch apart) 1/8-inch holes in a circular pattern in an aluminum tube. The length of this tube will be determined by how large you feel your silencer needs to be, which is a factor of the caliber and type of firearm you're trying to quiet.

The idea is that this tube must fit over the barrel of the gun, plus hold bottle caps, then fit inside an outer sleeve of plastic. I suggest 6 inches of 1 1/2-inch aluminum pipe.

Take a selection of plastic caps from 2-liter soft drink bottles and carefully drill out the centers to the caliber of your gun. Using an alignment rod that matches that caliber, slip the drilled caps over the rod and into the aluminum tube.

Carefully wrap this cap-filled inner tube with steel wool, fiberglass insulation, or metallic sponge material. Wrap it until it will fit snugly inside the outer tube of the silencer, usually a 7-inch piece of 2-inch plastic or aluminum pipe.

You need an end cap, aluminum or plastic with a center hole drilled to the firearm caliber, to close the one end of the

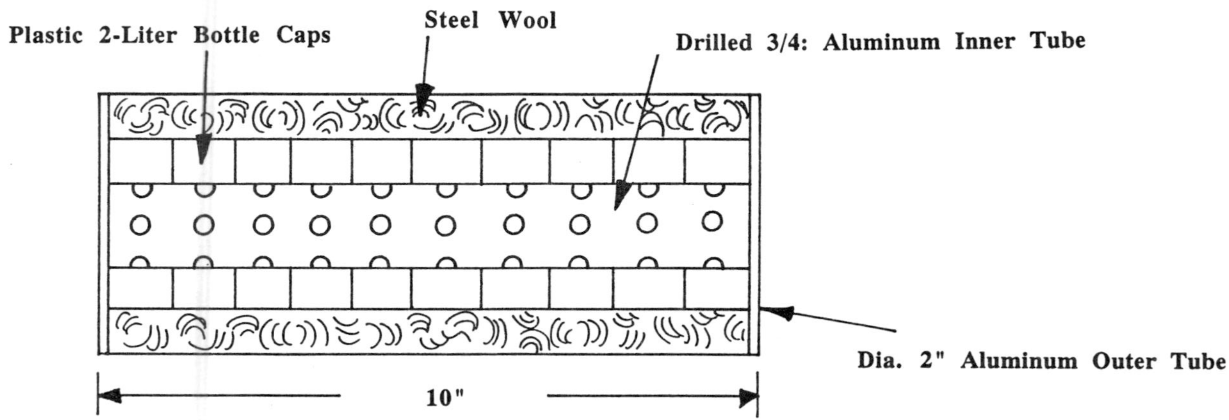

Figure 3. The Quirk Can is an amalgamation of various designs and is very easy to assemble. It has been field-tested by its designer, Thomas D. Quirk III. Credit: Donna Musick.

unit. The other end can be a screw, clamp, or compression fitting such as those used with other designs. What you use is dependent upon your firearm and whether you've threaded the barrel.

Mr. Quirk has seen this improvised silencer design used both with .22- and .25-calibers with very satisfactory sound-damping results.

As another tip, if you're threading the barrel, Mr. Quirk says to carefully glue a piece of felt paper, with the felt side facing inward, to the inside of that back tube opening. This way, the felt protects the barrel from being scratched when you fasten the unit to your firearm.

8 Rev Gene's Recycled Regulator

As promised, here is Rev Gene's environmentally correct can for the '90s. Conceived in the tightest government-secured research facilities imaginable, i.e., a prison, this design is being shown to the public for the first time on these pages. It's also a truly recycled effort, using commonly scrounged materials. The touchy-feely sorts call these "found materials of nature."

This is a silencer that's built from the inside up because of the design's nature. It all begins with a 6-inch length of metal screening, either hardware cloth or porch screen, rolled and formed around a pencil or dowel. You need about two or three thicknesses.

Before you begin this next step, keep in mind that the entire design needs to fit inside a heavy rubber hose, e.g., the radiator hose from a car, truck, or tractor. Of course, you can use a heavier, wider-diameter hose to increase the silencer's effectiveness, but beyond a certain point of size/silencing ratio, it gets counterproductive. For example, during the Vietnam unpleasantness, some brilliant scholar in our military was trying to design a silencer for the GE 7.62mm minigun. The design required to reduce the sound by even 10 percent was just slightly smaller than a 55-gallon drum. Point taken? That's a true story, by the way.

Sorry, I digressed. You wrap the next course around this

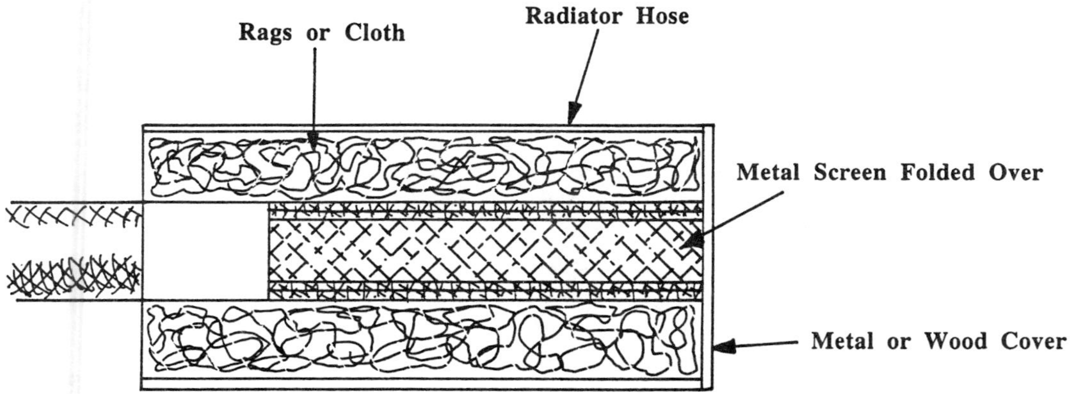

Figure 4. Rev Gene says this one was put together with ideas, tools, and materials easily obtained inside a prison. So much for the Clinton Crime Bill lie about gun control. Credit: The Bellisario Choir and Chorus.

central core of screening, choosing your material from among the following: home insulation, ceiling tile filler, heavy bath towel material, cotton rags, cloth, etc. Don't use synthetic fabrics, though, as they tend to flame or melt.

Make sure that as you add these wrappings you keep them as tight as possible against the screening, without collapsing it. Keeping the dowel piece or pencil in place will assure structural alignment integrity, of course.

The final step is to slide all of this inside a length of rubber hose about 6 1/2 inches long. Using Miracle Glue or some other instant, on-for-life modern stickum, mount a rubber washer to the one end of the hose, pushing the inner core snug against the washer walls. Make absolutely certain that the doweling or pencil can be easily pushed through the hole in the washer. We gun gurus call that positive alignment. Cough.

To mount Rev Gene's can on your gun, simply slide the open end of the hose silencer onto the barrel of the gun, being certain that it lines up with the core of the screening. That's the other end of alignment, the object being that the bullet passes through the barrel and both ends of the silencer, touching as little of either's innards as possible.

Finally, you can secure it with pressure screws mounted in a tripod fashion through the rubber and against the gun's barrel, or use clamps or tape, depending on your personal degree of upscaleness.

9

Fred G. Snopes Bag Lady's Special

Fred G. Snopes was formerly a gentried land owner and influential man of Southern means. But he fell afoul of the impatient wrath of Uncle David and Dollar Bill for not living up to his post-retirement military obligations according to the bylaws of the American Association of Oldepharts.

Thus, Fred and his 32nd wife, the former Bobbi Abigail Diptwick, age 12, were forced to homestead on the family's landfill, where Fred foraged for rats to feed his growing brood of little ones. It was here, on the landfill, that Fred designed the silencer that bears his name, just as Maxims did for his designs.

To recreate the Snopes design in your very own home or workshop, you'll need some sponges, newspaper, bath towel, heavy cardboard, the inner cardboard tube from a paper towel roll, electrical tape, stick-for-life glue, and some screws.

The first step is to pack sections of the sponge inside an 8- or 9-inch length of the cardboard tube. Pack about 3 inches, then insert a circular piece of the heavy cardboard cut to fit the inner diameter of the tube, i.e., a cardboard washer without the hole.

Repeat that step three more times until you have filled the tube to about a half-inch from the end. Glue a double set of heavy cardboard circles to the front end of the silencer, cut to fit the outer diameter.

Wrap several thicknesses of newspaper around this inner

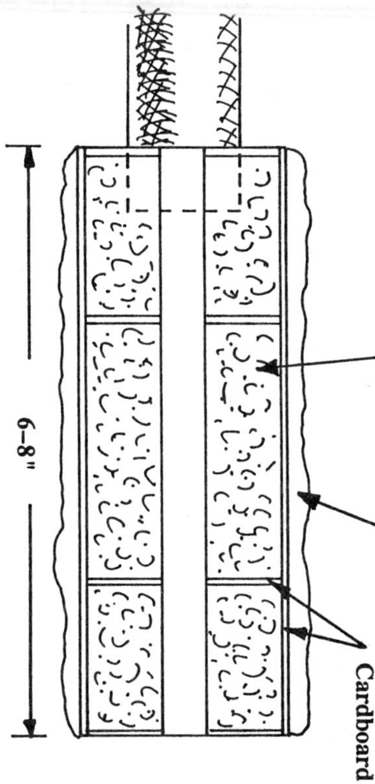

Figure 5. Conceived with devastating simplicity, a gift of its designer, the Snopes Bag Lady Special lays waste the most annoying noises. Credit: The Joyce Vincent Serendipity Center.

tube of cardboard. Fasten with electrical tape, going around the length of the tube several times. You want about a half-newspaper thickness.

Next, wrap the unit with a piece of the bath towel, being sure to cut all pieces that do not encircle neatly. Overlap, but don't cover that front end piece of cardboard. Wrap several layers of electrical tape around this unit for stability.

Take an electric or hand drill and, using a 1/4- or 3/8-inch bit, bore carefully through the end of the central part of the silencer mass. Use a small vise to maintain steady alignment. To mount the unit to your firearm, simply slide the barrel through to the drilled core until you feel it engage. Then fasten the unit to your firearm.

If you want to stabilize the unit and the gun, use three machine screws in tripod fashion through the silencer and onto the gun barrel. Tighten to fit and hold. Or, if you don't want to use screws, clamps or electrical tape will hold the unit on your gun.

10
Clay Shaw's Warped Wastrel

Not every operation can be planned, executed, and covered with total success all of the time, even with the best executive actions aiming for the top. There must be expedients, as the former clandestine military and espionage operative Clay Shaw knew so well. That's why he passed along this expedient silencer design to be shared with you.

His materials list starts with some sort of aluminum or plastic outer container, e.g., a soft drink can, shampoo bottle, etc. You'll also need a 6-inch length of steel pipe nipple. The diameter depends upon the caliber of silencer you're building, e.g., 1/8 inch for .22; 1/4 inch for 9mm, 7.62mm, or .380; and 3/8 inch for .45 ACP. You also need two steel pipe couplings to match your pipe nipple, copper mesh sponges, and about 6 inches of 1- to 1 1/2-inch cotton cloth.

The initial construction step, according to Clay, is to drill a hole in the outer container, sized to match the outside diameter of the pipe nipple. Next, drill four rows of bleed holes in the pipe nipple. (See the chart on page 56 for diameter and location.)

Then you thread one of the couplings on the pipe nipple. Now, pack the copper mesh sponges around the pipe nipple, getting them snug, but not squashed. Wrap the cotton cloth around the copper and secure with electrical tape.

The final step is to secure this core unit inside the outer

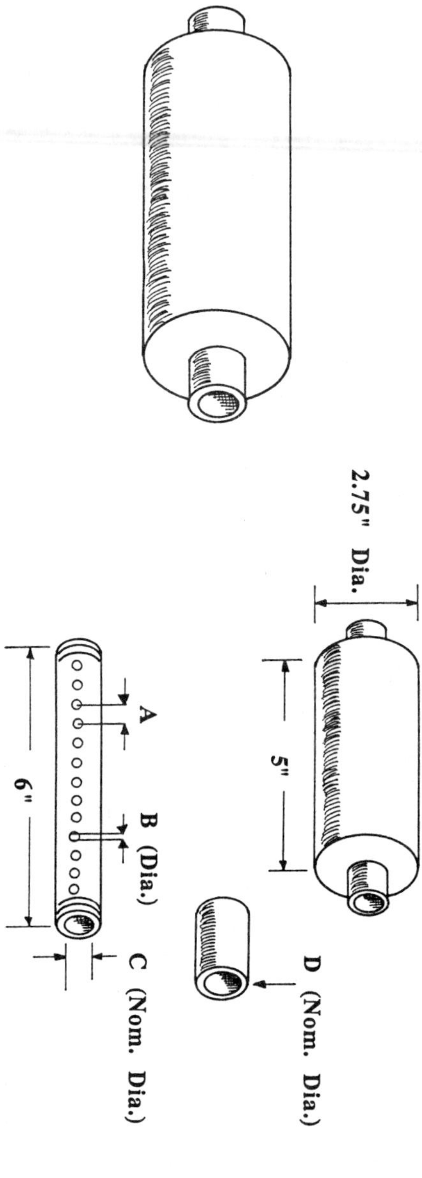

Figures 6, 7, 8. Key components for the Warped Wastrel silencer are easily found and assembled materials that are common to hardware stores and home workshops. Assembly is simple also. Credit: Hitters 'R' Us, courtesy of Clay Bertrand.

Clay Shaw's Warped Wastrel

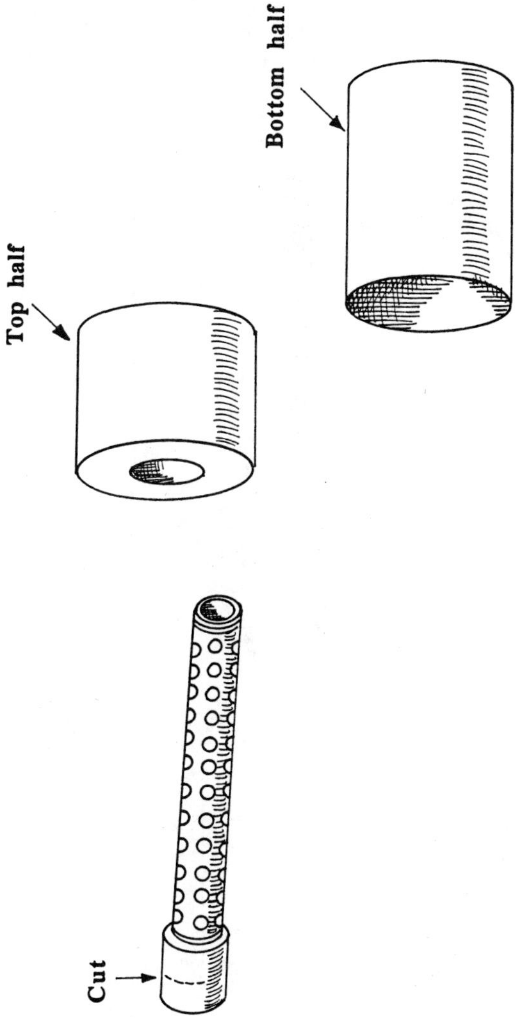

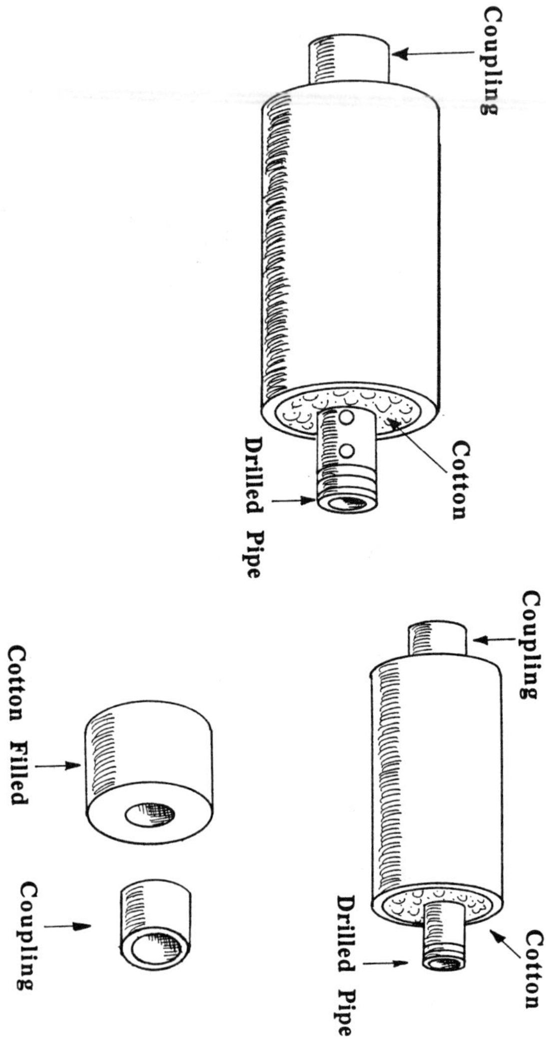
Coupling
Cotton
Drilled Pipe
Coupling
Cotton
Drilled Pipe
Cotton Filled
Coupling

container. Now thread the other coupling onto the other end of the pipe nipple.

To mount this silencer to your gun, you can either have the barrel threaded to match the pipe coupling and simply screw it on, or you can butt the barrel tight against the coupling and secure it with a compression nut or with screws set in tripod fashion through the coupling and onto your gun's barrel. The two important factors are alignment, of course, and making the connection as snug as possible to contain escaping gases.

It's a good design from the man who sprayed fear like a skunk sprays piss.

11

Queen Victoria's Venomous Vibraphone

Thanks to this good friend, a merry musical master, I can finally share Victoria's secret with you, an 8-inch, .22-caliber silencer that changes your gun's BANG to a piddling *plutt*.

A true romantic and every inch the Good Lady, Queen Victoria asks only that good deeds come from the actual use of her formerly secret silencer design. Being the disciplined humanitarian that I have always been, I agreed to share her design with that caveat in mind.

You start with ten 1-inch outer diameter (OD) aluminum washers with 1/4-inch centers. You will also need two 1-inch OD rubber stoppers, a package of Chore Boy/Girl copper sponge, a threaded end cap, and a threaded tube. The final item may be purchased legally, at least as this book is written. If they are no longer available for sale when you're reading this, you'll have to make a new friend, i.e, an honest, top-quality machinist who can make parts for you legally once you have your approved BATF Form 1.

Assembly is around a 1/4-inch dowel, which is .22-caliber size. At the muzzle end of the silencer, start with rolled or donut-formed units of the copper sponge, rolling them around the dowel. The thickness needs to be 1 inch OD around the dowel.

Slide an aluminum washer on each end of the copper sponge to compact it somewhat. This will serve as your silencer's expansion chamber.

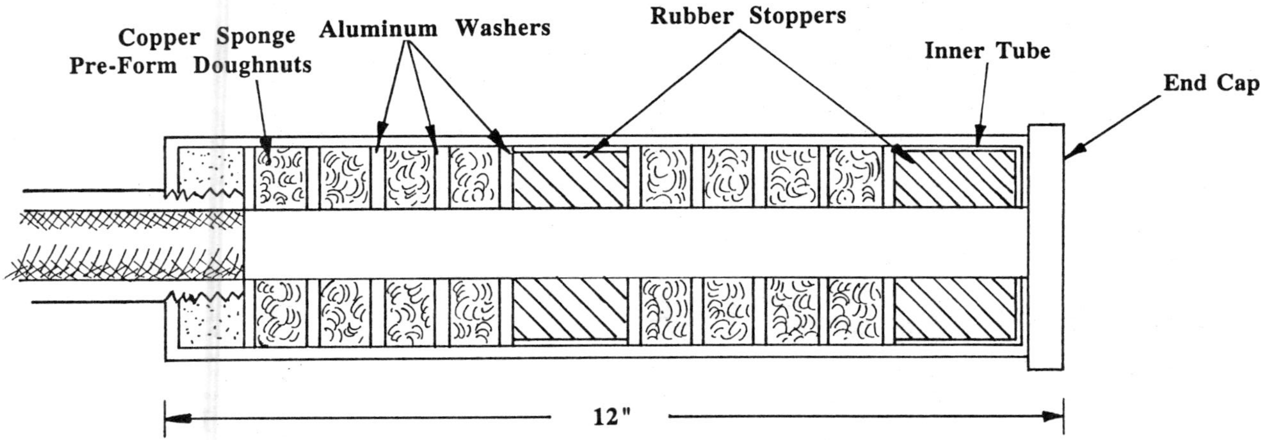

Figure 9. The very efficient Venomous Vibraphone is not an easy silencer to create and is based on a very established design principle already in use by at least four covert ops outfits in the U.S. and in the Middle East. Enjoy . . . Credit: Delhi Roo Enterprises.

Next, slide the washers along the dowel, slipping a rubber stopper between each group of four washers. Thread the end cap on the tube.

Carefully slide the inner unit that you've assembled around the dowel into the outer tube. Be sure that you align the dowel so it sticks out the end cap.

Finally, seal the entire unit by attaching the barrel adapter unit to the other end of the outer tube. You can mount this homemade silencer to your pistol or rifle by threading the weapon's barrel, or you can buy one of the many slip-on or clamp-on adapters sold at gun shows and through gun publications.

A word of advice from the nonvolatile Victoria—if you're going to thread your gun's barrel, purchase a spare barrel and have it threaded so that you always have the option of equipping your gun with an unthreaded barrel. This can be for aesthetic, image, legal, or investment reasons.

12

Dilligaf: The Sarah Brady Prevaricator Special

When I get nostalgic about my youth and my formerly free nation, I worry whether our younger generations will have the balls and brains to carry out the coming revolution that we need to recapture our freedom. They'd better hurry before we lose the rest of our Bill of Rights in the same way the Second Amendment is dying of insidious cancer of the misguided in body Amerika.

In these scary times of dictatorship, I am reminded of the too-true words of Thomas Jefferson, who wrote, "The tree of liberty must be refreshed from time to time with the blood of patriots and tyrants. It is its natural manure."

Speaking of manure, though, always puts me in mind of Sarah Brady and her storm troopers from the CIA-founded Handgun Control Fascists. It's quite ironic that Omar Ahmed "Skippy" Blufarrb has dedicated his DILLIGAF silencer design to Sarah Brady.

"It's because I would like to put a silencer on the lies that come out of that evil witch's mouth as she tries to take away your hard-won American freedoms," Skippy said.

Skippy's design will work best with a .22-caliber firearm, so let's get right to the engineering specifics. You need an 8- to 9-inch-long tube of 1-inch OD heavy-grade aluminum. You also need the fillings for the tube; Skippy calls them the can's innards. He suggests a combination of three metal baffles,

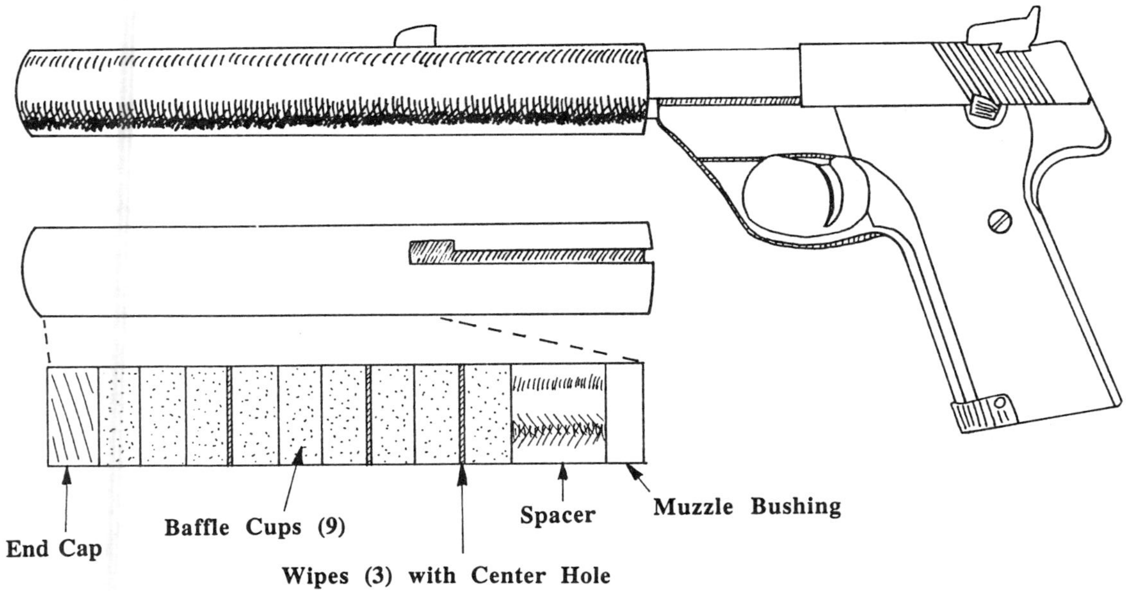

Figure 10. The Dilligaf design is adaptable to almost any firearm. I've shown it here with a Hi Standard HD .22 pistol, using an outer sleeve cutout to match the sight of the original firearm we used for testing. It can be attached without the cutout by other methods if the firearm's front side is removed. Credit: Tim Powala.

two washers or bronze sponge donuts, followed by three metal baffles.

You will need a nominal 3/8-inch hole in each baffle for the round to pass through. Don't worry if it's not exact. The main concern is that the alignment be precise; more on that later.

You need wipes, and Skippy suggests rubber disks or rubber washers, some with 1/4-inch centered holes, and some with X-cross-cuts. You will need about 1/4- to 3/8-inch of wipes at the muzzle end of the can, just behind the end cap.

You place the center-holed wipes next to the last course of baffling, followed by the X-cross-cut wipes last. The best mix is to have three holed wipes followed by three of the X'd ones.

Assemble the outer tube using the end cap method described in other chapters, i.e., with a dowel rod, working from muzzle end to the outer end. The last internal item is the 1/4 inch of rubber wipe(s). Then simply attach the end cap, either threaded or compression, your choice.

Attaching the Sarah silencer to your favorite .22-caliber firearm is a matter of individual adjustment to that particular item. For example, you may wish to use a duplicate barrel so you don't have to alter or mar your existing one. The same instructions apply to the installation of the Sarah model as to others mentioned previously.

13

Quieting Blowhard Billary

What made you think this was something new, just because it was named after the yuppie generation's first Co-President? Actually, this is just a refinement of the old soft drink suppressor made famous by James B. Adair and included in my first book on this arcane topic.

Several readers who took active interest in the initial design for a soda pop silencer, i.e., one using a 2- or 3-liter plastic jug for the body of the unit, included Raul Pindy, Flame Blackwell, and Thurston Whyst.

One of the Blackwell suggestions was to fill the soft drink container about half full of that expandable plastic foam, being sure to insert a carefully centered greased dowel rod of the appropriate size to the caliber of the firearm you plan to mount with the unit. This dowel is removed when the foam hardens so that you have proper barrel alignment.

Another of Blackwell's more erotic-sounding suggestions was to fill the container with a large, moist loofah sponge as a sound and thermal absorbent.

Both Pindy and Whyst agreed on some sort of sponge filler, but Whyst added the idea that you should spray-paint the outside of the plastic container with auto primer. He thought this would increase the life of the unit, especially if you were to glue several sections of

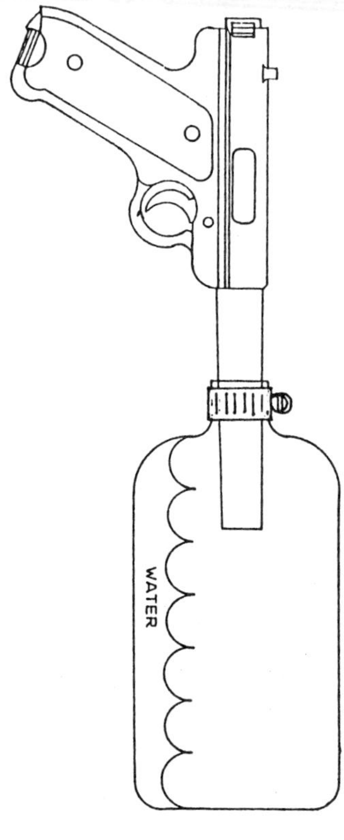

Figure 11. As Jim Adair notes, "I'm a pepper and you're dead," which is a rather terminal view of this fun silencer made from a two-liter pop bottle. Several experts in this arcane field suggest filling the bottle with loofah sponge and/or other audiological dampers before application. This unit remains one of the cheapest, simplest, and easiest improvised silencers on the market. Credit: Melissa Kate and Grover.

rubber or plastic to the "muzzle" end of the multiliter container.

If only we could shut up and move out our Co-Presidents as effectively and quietly back to private life.

14
The Can Can

More proof positive about the impossibility of gun control, here is yet another silencer that came from a prison mastermind. Is there any end to it? The Merry Mormon passed this design along, saying it was the collaborative work of Michel Mertz, an infamously coarse union man, and his associate, Q.J. Win.

Obviously an improvised design, the Can Can is a combination of newspapers, corrugated cardboard, and metal washers. The finished product ranges from 6 to 8 inches long and is as thick as the maker rolls it.

Space three or four washers evenly along a dowel rod or pencil the size of your firearm's caliber. A normal-sized pencil works well for a .22; a carpenter's pencil or 5/8-inch dowel will work for a 9mm or .380. The washers form the inner structural framing for your Can Can.

Lay the washered dowel along a four- or five-page section of newspaper that's folded to a 6- to 8-inch length. Carefully attach a sleeve extension of corrugated cardboard to the end of the newspaper to act as a barrel mount. You can use parcel strapping tape to attach the cardboard to the newspaper.

Then carefully roll the paper/cardboard around the structural core. When you're finished, you should have several layers of newspaper, at least 1/2-inch thick, around the core. Use the strapping tape to secure it. Then circle this unit with a layer of corrugated cardboard to act as an outer tube.

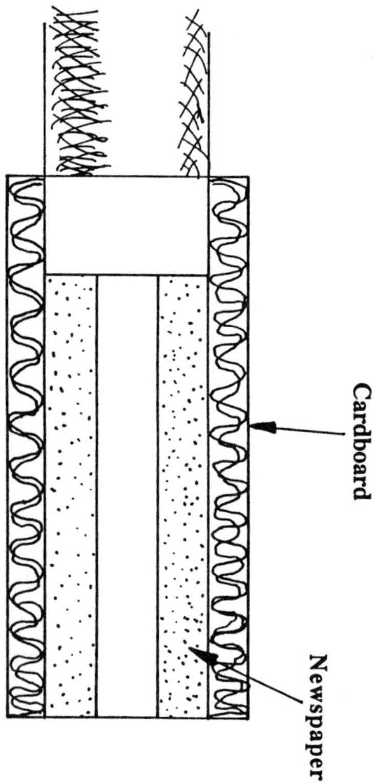

Figure 12. Another in-house specialty, hence its name, the Can Can was easily designed and built, and, according to firsthand observation, worked perfectly. Credit: The Mormon Archives.

Run strapping tape around the entire unit to secure it further. You need to glue or tape a circular piece of corrugated cardboard over the newspaper portion of the Can Can as a front end cap. You are now ready to mount the unit from the other end, i.e., the end with the cardboard sleeve.

First, remove the dowel/pencil and washers. You may have to pack additional cardboard or use washcloth material to securely mount the unit from the cardboard end to your firearm's barrel. The barrel should insert about 2 inches into the cardboard, then into the Can Can's body.

Secure the unit to your firearm with strapping or electrical tape. Before you fire the firearm, use the dowel rod to be certain that there is proper alignment between the gun's barrel and the silencer.

There you have it. And, just between us, when you lip read, be careful you don't get ink on your mouth.

15 Tool Time Can

Having a tool man as a friend greatly expedites the production of your new homemade silencer. As you've probably seen by now, building that silencer is a great deal like the comparison between sex and the game of bridge—you have to have a good partner or a good hand.

With that in mind, here is a home workshop special using parts found in every handy tool man's home. We start with the silencer's outer body, an empty grease gun cartridge, like the Binford Tools Mk 602, an 8" x 3" cylinder.

Remove the cartridge's ends and set aside. Locate a small plastic plant cage like the ladies of garden clubs use for their little posies and set that aside. Do the same thing with some steel wool or metal sponge.

Next, secure your rifle or pistol's barrel in a vise and drill two rows of six or seven 1/8-inch holes in alternating fashion on each side of the barrel (see Figure 13, p. 78). You will have drilled four rows in all that way to port the barrel.

Insert the plastic plant cage over the barrel as an improvised gas and heat chamber. The next step is to either wrap the metal sponge around the plant cage or stuff the steel wool in and around the cage, effectively covering the ported barrel.

By now, as a veteran can builder, you fully understand the why and where this design is heading. Now comes the tight fit tricky part. You need to insert the grease gun cartridge over the inner packing covering your firearm's barrel.

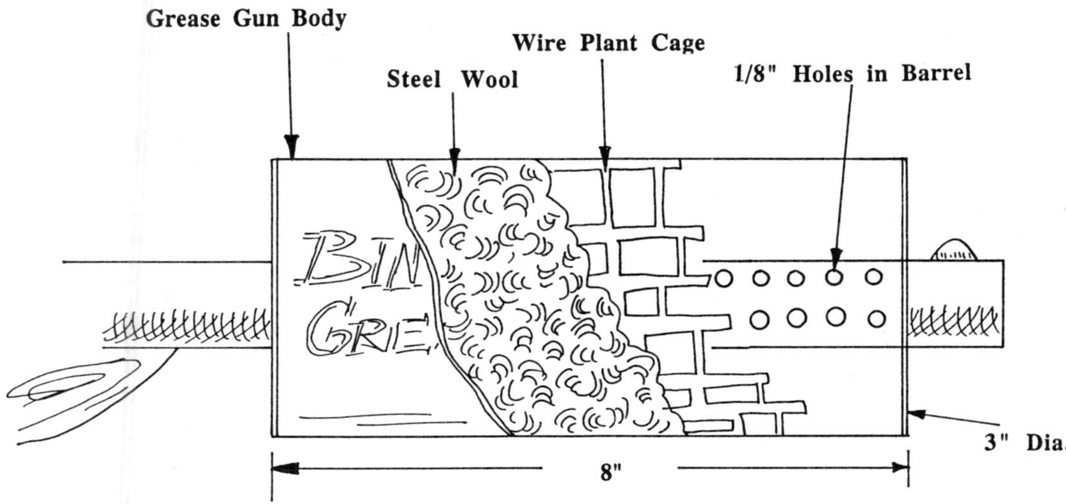

Figure 13. The Tool Time design combines the artistry of the workshop and the garden club for its muffled motif. It's a must for your next in-house project. Credit: Vic and Jim of Capt Media Partners.

Obviously, you may need to modify the packing/wrapping of the barrel. That can be done easily. The main point is that the cartridge has to slide over the inner unit, flush to the receiver. If it is not flush you will need to either seal it with tape or, if you wish to be more fancy, build an end cap just as I suggested with other designs.

You will need a muzzle end cap, of course. I suggest using wipeless designs, but you're welcome to use a rubber stopper or something similar if you remember to include an adequate exit hole for the .22-caliber round. Also, you may wish to check the orientation of the barrel and end cap before firing, using the ubiquitous piece of dowel rod.

This is an easy silencer to build, with or without a tool man's help. Just remember what tool man Kevin Stiffler said as he examined his bat: "I was going to read Rev. Peale's book, *The Power of Positive Thinking*, but what the hell good would that do?"

Appendix

STANDING ORDERS, ROGERS' RANGERS (MAJ. ROBERT ROGERS, 1759)

1. Don't forget nothing.
2. Have your musket clean as a whistle, hatchet scoured, 60 rounds powder and ball, and be ready to march at a minute's warning.
3. When you're on the march, act the way you would if you was sneaking up on a deer. See the enemy first.
4. Tell the truth about what you see and what you do. There is an army depending on us for correct information. You can lie all you please when you tell other folks about the Rangers, but don't never lie to a Ranger or officer.
5. Don't never take a chance you don't have to.
6. When we're on the march, we march single file, far enough apart so one shot can't go through two men.
7. If we strike swamps or soft ground, we spread out abreast, so it's hard to track us.
8. When we march, we keep moving till dark, so as to give the enemy the least possible chance at us.
9. When we camp, half the party stays awake while the other half sleeps.
10. If we take prisoners, we keep 'em separate till we have

had time to examine them, so they can't cook up a story between 'em.

11. Don't ever march home the same way. Take a different route so you won't be ambushed.
12. No matter whether we travel in big parties or little ones, each party has to keep a scout 20 yards ahead, 20 yards on each flank, and 20 yards in the rear, so the main body can't be surprised and wiped out.
13. Every night you'll be told where to meet if surrounded by a superior force.
14. Don't sit down to eat without posting sentries.
15. Don't sleep beyond dawn. Dawn's when the French and Indians attack.
16. Don't cross a river by a regular ford.
17. If somebody's trailing you, make a circle, come back onto your own tracks, and ambush the folks that aim to ambush you.
18. Don't stand up when the enemy's coming against you. Kneel down. Hide behind a tree.
19. Let the enemy come till he's almost close enough to touch. Then let him have it and jump out and finish him up with your hatchet.

Real Books Written by the Real Experts

Although a few of these books are out of print, you can sometimes find used copies at gun shows, flea markets, used book stores, and from book dealers who advertise that they'll find used books for you. Check ads in *Gun List* and *Shotgun News*. Or look for a used book research company in a city telephone directory.

Dater, Dr. Philip H. *The Art of Silence*. Boise, ID: Antares Technologies, 1982.

Flores, J. *How to Make Disposable Silencers*. Rosemead, CA: J. Flores Publications, 1985.

______. *How to Make Disposable Silencers*, Vol. II. Rosemead, CA: J. Flores, 1985.

Home Workshop Silencers I. Boulder, CO: Paladin Press, 1980.

Huebner, Siegfried F. *Silencers for Hand Firearms*. Boulder, CO: Paladin Press, 1976.

McLean, Donald B. *Firearms Silencers, Vol. I*. Forest Grove, OR: Normount Technical Publications, 1968.

Minnery, John A. *Firearms Silencers, Vol. II*. Cornville, AZ: Desert Publications, 1981.

Nye, Roy D. *The Handy Dandy, Super Duper Junkyard Silencer Book*. Valley Farms, AZ: Roy D. Nye, 1986.

Santini, A.Z. and Chuck Carson. *The Silencer*. Tucson, AZ: The A&L Publishing Co., 1980.

Thomas, Donald G. *European Silencer Patents*. Boulder, CO: Paladin Press, 1978.

_____. *U.S. Silencer Patents*. Boulder, CO: Paladin Press, 1973.

Truby, J.David. *Silencers, Snipers & Assassins*. Boulder, CO: Paladin Press, 1972; second edition, 1987.

_____. *Quiet Killers*. Boulder, CO: Paladin Press, 1972, 1979.

_____. *Modern Firearm Silencers*. Boulder, CO: Paladin Press, 1983, 1992.

Wells, Robert. *The Anarchist Handbook*. Miami: J. Flores Publications, 1983.

_____. *The Silencer Cookbook*. Cornville, AZ: Firepower Publications, 1983.

•••

With apologies to the real experts of suppressorland . . .

Jonathan Arthur Ciener
Dr. Philip H. Dater
Charles Erb
Mickey Finn
Reed Knight
John Norrell
James Ryan
James Wilson

•••

***With your help we'll have Volume* III . . .**

If any of you have a good silencer design, send me your sketch and description. If I use the design in Volume III, I'll buy you a free book, autograph it, and personally send it to you.

George W. Hayduke
P.O. Box 1307
Boulder, CO 80306